# FUNDAMENTALS
# OF SENSORS
# FOR ENGINEERING
# AND SCIENCE

# FUNDAMENTALS
# OF SENSORS
# FOR ENGINEERING
# AND SCIENCE

**PATRICK F DUNN**
University of Notre Dame
Indiana, USA

**CRC Press**
Taylor & Francis Group
Boca Raton   London   New York

CRC Press is an imprint of the
Taylor & Francis Group, an **informa** business

CRC Press
Taylor & Francis Group
6000 Broken Sound Parkway NW, Suite 300
Boca Raton, FL 33487-2742

International Standard Book Number: 978-1-4398-6103-5 (Paperback)

**Visit the Taylor & Francis Web site at**
**http://www.taylorandfrancis.com**

**and the CRC Press Web site at**
**http://www.crcpress.com**

# Cover

The two sensors featured on the cover are span temperature measurement chronology. The thermoscope marks its beginning. A modern temperature-field sensor symbolizes the present.

The two-bulb "Dutch" thermoscope is depicted on the top left of the cover. This is just one variation of the thermoscope, which was invented in Europe around the turn of the 16th century. Who is its inventor is debatable. The thermoscope was invented about 40 years before the first sealed thermometer and more than 100 years before Gabriel Fahrenheit developed his temperature scale. The thermoscope indicated changes of temperature but did not measure them. Its sensor was the liquid within the thermoscope, which usually was water.

The grid of red laser beams shown on the bottom right of the cover is produced by an arrangement of fiber optics used for gas temperature imaging. The arrangement has been developed by Professor Scott Sanders's research group at the University of Wisconsin—Madison. For the photo shown, a visible laser source was coupled into each of the 30 delivery fibers, and the measurement plane was visualized using fog. In actual operation, a specially designed infrared laser source is coupled into the fibers. Water vapor in the measurement plane absorbs some of this infrared radiation, and, by analyzing the absorption, gas temperatures are determined. Tomographic reconstruction is used to generate planar gas temperature images from the 30-beam absorption data. The apparatus can provide movies of gas temperature with a frame rate of 50 kHz (20 ms/frame) and accommodates gas temperatures up to 2500 K.

# Contents

# Preface

This text describes the fundamental aspects of sensors that currently are used by both engineers and scientists throughout the world. Its material is presented in a contemporary style to enable the user to make informed decisions when choosing a sensor. Forty-five different man-made sensors are considered. In addition, the sensors of the human body are reviewed. Finally, several biomimetic sensors that mimic their human counterparts are presented. Their examples exemplify the current trend toward making sensors smaller, more precise, and more robust.

Besides serving as a text on sensor fundamentals, this text is a companion to the second edition of **Measurement and Data Analysis for Engineering and Science**. This new text greatly expands the coverage of sensors presented in the second edition's Chapter 3 (Measurement Systems). This project grew out of requests by instructors to increase coverage of modern sensors and their basic principles.

This text follows a unique approach. It discusses the role of a sensor, its characteristics, and the various ways in which it is classified. Contemporary sensors are organized and described with respect to their basic physical principles. A new feature is coverage of human sensors, which are the ultimate goal of many biomimetic sensor designers. Several recent biomimetic sensors are described to illustrate recent progress in biomimetic sensor design.

This text's web site (www.nd.edu/~pdunn/www.text/sensors.html) should be consulted for up-to-date information, as well as that for the second edition of **Measurement and Data Analysis for Engineering and Science** (www.nd.edu/~pdunn/www.text/measurements.html). Instructors who adopt either text for their course can receive a CD containing the particular problem solutions manual by contacting their Taylor & Francis / CRC Press representative.

Many people contributed to the two editions of **Measurement and Data Analysis for Engineering and Science** and, thus, to this new text. They are acknowledged in those editions. Two individuals that have contributed notably to both texts deserve special mention. They are Jonathan Plant, the editor of both editions of my measurements text, who suggested writing this new text, and my wife, Carol, who has supported me all along the way.

Patrick F. Dunn
University of Notre Dame, Notre Dame, Indiana

# *Author*

**Patrick F. Dunn, Ph.D., P.E.**, is a professor of aerospace and mechanical engineering at the University of Notre Dame, where he has been a faculty member since 1985. Prior to 1985, he was a mechanical engineer at Argonne National Laboratory from 1976 to 1985 and a postdoctoral fellow at Duke University from 1974 to 1976. He earned his B.S., M.S., and Ph.D. degrees in engineering from Purdue University (1970, 1971, and 1974). He is the author of more than 160 scientific journal and refereed symposia publications and a licensed professional engineer in Indiana and Illinois. He is a Fellow of the American Society of Mechanical Engineers and an Associate Fellow of the American Institute of Aeronautics and Astronautics. He is the recipient of departmental, college, and university teaching awards.

Professor Dunn's scientific expertise is in fluid mechanics and microparticle behavior in flows. He is an experimentalist with more than 40 years of experience. He is the author of the textbook **Measurement and Data Analysis for Engineering and Science** (first edition by McGraw-Hill, 2005; second edition by Taylor & Francis / CRC Press, 2010) and **Uncertainty Analysis for Forensic Science** with R.M. Brach (first and second editions by Lawyers & Judges Publishing Company, 2004 and 2009).

## UNIT CONVERSIONS

| | | | | | | |
|---|---|---|---|---|---|---|
| **Length** | 1 m = | 100 cm | $1 \times 10^{-3}$ km | 39.37 in. | 3.281 ft | $6.214 \times 10^{-4}$ mi | $3.937 \times 10^4$ mil |
| **Area** | 1 m² = | $1 \times 10^4$ cm² | 10.76 ft² | 1550 in.² | $2.471 \times 10^{-4}$ acre | $1 \times 10^{-4}$ ha | $3.861 \times 10^{-7}$ mi² |
| **Volume** | 1 m³ = | $1 \times 10^6$ cm³ | 1000 L | 35.31 ft³ | $6.102 \times 10^4$ in.³ | 264.17 US gallon | 1056.7 liquid qt |
| **Time** | 1 s = | 1000 ms | $1.667 \times 10^{-2}$ min | $2.778 \times 10^{-4}$ h | $1.157 \times 10^{-5}$ d | $3.169 \times 10^{-8}$ y | $3 \times 10^8$ light m |
| **Speed** | 1 m/s = | 100 cm/s | 3.281 ft/s | 3.6 km/h | 2.237 mi/h | 1.944 nautical mi/h | |
| **Mass** | 1 kg = | 1000 g | $6.852 \times 10^{-2}$ slug | 2.2046 lbm | $1 \times 10^{-3}$ metric ton | $6.023 \times 10^{26}$ amu | 500 carat |
| **Mass Density** | 1 kg/m³ = | 0.001 g/cm³ | $1.940 \times 10^{-3}$ slug/ft³ | $6.242 \times 10^{-2}$ lbm/ft³ | $1.123 \times 10^{-6}$ slug/in.³ | $3.612 \times 10^{-5}$ lbm/in.³ | |
| **Weight, Force** | 1 N = | $1 \times 10^5$ dyne | 0.2248 lbf | 7.233 pdl | | | |
| **Pressure** | 1 Pa = | 10 dyne/cm² | $9.869 \times 10^{-6}$ atm | $4.015 \times 10^{-3}$ in. H$_2$O | $7.501 \times 10^{-4}$ cm Hg | $1.450 \times 10^{-4}$ lbf/in.² | $2.089 \times 10^{-2}$ lbf/ft² |
| **Energy, Work** | 1 J = | $2.778 \times 10^{-7}$ kW·h | $9.481 \times 10^{-4}$ Btu | $1 \times 10^7$ erg | 0.7376 ft·lbf | $3.725 \times 10^{-7}$ hp·h | 0.2389 cal |
| **Power** | 1 W = | 0.001 kW | 3.413 Btu/h | 0.7376 ft·lbf/s | $1.341 \times 10^{-3}$ hp | 0.2389 cal/s | |
| **Temperature** | 1 K = | $(5/9) \times$ °F + 255.38 | °C + 273.15 | $(5/9) \times$ °R | | | |
| **Plane Angle** | 1 rad = | 57.30° | 3438' | $2.063 \times 10^5$" | 0.1592 rev | | |

## SI DERIVED UNITS EXPRESSED IN BASE UNITS

| Derived Unit | Symbol | Base Units |
|---|---|---|
| Force | N (newton) | $kg \cdot m \cdot s^{-2}$ |
| Pressure | Pa (pascal) | $kg \cdot m^{-1} \cdot s^{-2}$ |
| Energy, Work, Heat | J (joule) | $kg \cdot m^2 \cdot s^{-2}$ |
| Power | W (watt) | $kg \cdot m^2 \cdot s^{-3}$ |
| Electric Charge | C (coulomb) | $A \cdot s$ |
| Electric Potential Difference | V (volt) | $kg \cdot m^2 \cdot s^{-3} \cdot A^{-1}$ |
| Electric Resistance | Ω (ohm) | $kg \cdot m^2 \cdot s^{-3} \cdot A^{-2}$ |
| Electric Conductance | S (siemens) | $kg^{-1} \cdot m^{-2} \cdot s^3 \cdot A^2$ |
| Electric Capacitance | F (farad) | $kg^{-1} \cdot m^{-2} \cdot s^4 \cdot A^2$ |
| Electric Inductance | H (henry) | $kg \cdot m^2 \cdot s^{-2} \cdot A^{-2}$ |
| Magnetic Flux | Wb (weber) | $kg \cdot m^2 \cdot s^{-2} \cdot A$ |

| Name | Symbol | Value |
|---|---|---|
| Avogadro's number | $N_a$ | $6.022\ 142 \times 10^{23}$ entities/mole |
| Boltzman's number | $k_B$ | $1.3805 \times 10^{-23}$ J/K |
| Elementary electrical charge | $e$ | $1.6022 \times 10^{-19}$ C |
| Faraday's constant | $F$ | $96\ 485$ C |
| Gravitational acceleration (standard) | $g_o$ | $9.806\ 65$ m/s$^2$ |
| Gravitational constant | $G$ | $6.673\ 00 \times 10^{-11}$ m$^3$/(kg·s$^2$) |
| Permeability of free space | $\mu_o$ | $4\pi \times 10^{-7}$ H/m |
| Permittivity of free space | $\epsilon_o$ | $8.8544 \times 10^{-12}$ F/m |
| Planck's constant | $\hbar$ | $6.6256 \times 10^{-34}$ J·s |
| Universal gas constant | $\mathcal{R}$ | $8313.3$ J/(kg-mole·K) |
| Velocity of light $in\ vacuo$ | $c$ | $299\ 792\ 458$ m/s |

PHYSICAL CONSTANTS

# 1

## Sensor Fundamentals

### CONTENTS

## 1.1   Chapter Overview

Sensors are at the beginning of every measurement system, whether it is a liquid in a thermometer responding to a change in temperature or a rod in our retina sensing a single photon of light. This chapter discusses the role of a sensor in a measurement system. Classifications of sensors are presented along with their characteristics. Additional considerations also are offered, including how sensor scaling affects its design, the instrument uncertainties of sensors, and sensor calibration.

## 1.2   Role in a Measurement System

A measurement system comprises the equipment used to sense an experiment's environment, to modify what is sensed into a recordable form, and to record its values. Formally, the elements of a measurement system include the sensor, the transducer, the signal conditioner, and the signal processor. These elements, acting in concert, sense the physical variable, provide a response in the form of a signal, condition the signal, process the signal, and store its value.

A measurement system's main purpose is to produce an accurate numerical value of the measurand. Ideally, the recorded value should be the exact value of the physical variable sensed by the measurement system. In practice, the perfect measurement system does not exist, nor is it needed. A result only

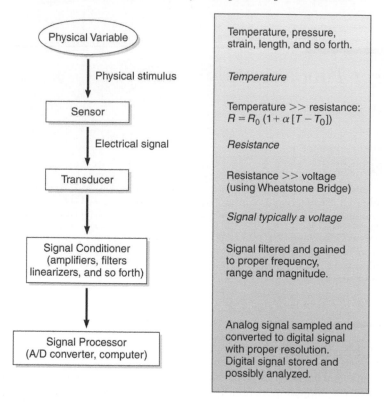

**FIGURE 1.1**
The general measurement system configuration.

needs to have a certain accuracy that is achieved using the most simple equipment and measurement strategy. This can be accomplished provided there is a good understanding of the system's response characteristics.

To accomplish the task of measurement, the system must perform several functions in a series. These are illustrated schematically in Figure 1.1. First, the physical variable must be sensed by the system. The variable's stimulus determines a specific state of the sensor's properties. Any detectable physical property of the sensor can serve as the sensor's **signal**. When this signal changes rapidly in time, it is referred to as an **impulse**. So, by definition, the **sensor** is a device that senses a physical stimulus and converts it into a signal. This signal usually is electrical, mechanical, or optical.

For example, as depicted by the words in italics in Figure 1.1, the temperature of a gas (the physical stimulus) results in an electrical resistance (the signal) of a resistance temperature device (RTD), a temperature sensor that is located in the gas. This is because the resistance of the RTD sensor (typically a fine platinum wire) is proportional to the change in temperature from a reference temperature. Thus, by measuring the RTD's resistance, the local temperature can be determined. In some situations, however, the signal may not be amenable to direct measurement. This requires that the signal be changed into a more appropriate form, which, in almost all circumstances, is electrical. Most of the sensors in our bodies have electrical outputs.

The device that changes (transduces) the signal into the desired quantity (be it electrical, mechanical, optical, or another form) is the **transducer**. In the most general sense, a transducer transforms energy from one form to another. Usually, the transducer's output is an electrical signal, such as a voltage or current. For the RTD example, this would be accomplished by having the RTD's sensor serve as one resistor in an electrical circuit (a Wheatstone bridge) that yields an output voltage proportional to the sensor's resistance. Often, either the word *sensor* or the word *transducer* is used to describe the combination of the actual sensor and transducer. A transducer also can change an input into an output providing motion. In this case, the transducer is called an **actuator**. Sometimes, the term *transducer* is considered to encompass both sensors and actuators [1]. So, it is important to clarify what someone specifically means when referring to a transducer.

The sensor/transducer system in a house thermostat basically consists of a metallic coil (the sensor) with a small glass capsule (the transducer) fixed to its top end. Inside the capsule is a small amount of mercury and two electrical contacts (one at the bottom and one at the top). When the thermostat's set temperature equals the desired room temperature, the mercury is at the bottom of the capsule such that no connection is made via the electrically conducting mercury and the two contacts. The furnace and its blower are off. As the room temperature decreases, the metallic coil contracts, thereby tilting the capsule and causing the mercury to close the connection between the two contacts. The capsule transduces the length change in the coil into a digital (on/off) signal.

Another type of sensor/transducer system is in a landline telephone mouthpiece. This consists of a diaphragm with coils housed inside a small magnet. There is one system for the mouthpiece and one for the earpiece. The diaphragm is the sensor. Its coils within the magnet's field are the transducer. Talking into the mouthpiece generates pressure waves, causing the diaphragm with its coils to move within the magnetic field. This induces a current in the coil, which is transmitted (after modification) to another telephone. When the current arrives at the earpiece, it flows through the coils of the earpiece's diaphragm inside the magnetic field and causes the diaphragm to move. This sets up pressure waves that strike a person's eardrum as sound. Newer phones use piezo-sensors/transducers that generate an electric current from applied pressure waves and, alternatively, pressure waves from an applied electric current. Today, most signals are digitally encoded for transmission either in optical pulses through fibers or in electromagnetic waves to and from satellites. Even with this new technology, the sensor still is a surface that moves, and the transducer still converts this movement into an electrical current.

Often after the signal has been transduced, its magnitude still may be too small or may contain unwanted electrical noise. In this case, the signal must be conditioned before it can be processed and recorded. In the signal conditioning stage, an amplifier may be used to increase the signal's amplitude, or a filter may be used to remove the electrical noise or some unwanted frequency content in the signal. The **signal conditioner**, in essence, puts the signal in its final form to be processed and recorded.

In most situations, the conditioner's output signal is analog (continuous in time), and the **signal processor** output is digital (discrete in time). So, in the signal processing stage, the signal must be converted from analog to digital. This is accomplished by adding an analog-to-digital (A/D) converter, which usually is contained within the computer that is used to record and store data. That computer also can be used to analyze the resulting data or to pass this information to another computer.

A standard glass-bulb thermometer contains all the elements of a measurement system. The sensor is actually the liquid within the bulb. As the temperature changes, the liquid volume changes, either expanding with an increase in temperature or contracting with a decrease in temperature. The transducer is the bulb of the thermometer. A change in the volume of the liquid inside the bulb leads to a mechanical displacement of the liquid because of the bulb's fixed volume. The stem of the thermometer is a signal conditioner that physically amplifies the liquid's displacement, and the scale on the stem is a signal processor that provides a recordable output.

## 1.3  Domains

There are a variety of ways by which a sensor can be categorized. Often, a sensor is located within the environment of interest. This type of sensor, which usually is mechanical or electrical, is considered an **invasive**, or *in situ*, sensor. Ideally, invasive sensors should not disturb the environment, which could alter the process under investigation. A sensor also can be located outside the environment. For example, an optical pyrometer senses temperature remotely. This is a **noninvasive** sensor.

Almost all of the signals between the sensor and the detectable output are electrical, mechanical, or optical. Electrical-based sensors and transducers can be **active** or **passive**. Active elements require an external power supply to produce a voltage or current output. The electrical elements of active electrical sensors are resistors, capacitors, or inductors. Passive elements require no external power supply. Their elements typically are either electromagnetic or thermoelectric. Mechanically based sensors and transducers usually use a secondary sensing element that provides an electrical output. Often the sensor and transducer are combined physically into one device.

Sensors in the human body convert a stimulus input into an electrical output. These sensors include those for vision, taste, smell, hearing, equilibrium, touch, temperature, nociception, and proprioception. Because energy is required to restore the potential difference to its potential that existed prior to stimulus application, these sensors would be characterized best as active electrical sensors.

Sensors can be categorized into **domains**, according to the type of physical variables that they sense [1], [2]. These domains and the sensed variables include

- chemical: chemical concentration, composition, and reaction rate;

- electrical: current, voltage, resistance, capacitance, inductance, and charge;

- magnetic: magnetic field intensity, flux density, and magnetization;

- mechanical: displacement or strain, level, position, velocity, acceleration, force, torque, pressure, and flow rate;

- radiant: electromagnetic wave intensity, wavelength, polarization, and phase; and

- thermal: temperature, heat, and heat flux.

Sensors also can be organized with respect to the physical basis of how they sense. These are

- electric,

- piezoresistive,

- fluid mechanic,

- optic,

- photoelastic,

- thermoelectric, and

- electrochemical.

This is the manner by which the sensors described in Chapter 2 are presented.

## 1.4    Characteristics

The characteristics of a sensor can include those related to the sensor's input and output, which are the sensor's

- operational bandwidth,

- magnitude and frequency response over that bandwidth,

- sensitivity,

- accuracy,

- resolution,

- response time,

- recovery time, and

- output type.

All of these, except for the last, relate to how well the sensor responds to a stimulus. These response characteristics are described in further detail in Chapter 4 of [3].

Further, there are characteristics that describe the sensor as a component of a measurement system. These are sensor

- selectivity;

- voltage or current supply requirements;

- physical dimensions, weight, and materials;

- environmental operating conditions (pressure, temperature, relative humidity, air purity, and radiation);

- additional signal conditioning requirements;

- operational complexity; and

- cost.

Different sensors from which to choose can be assigned level of priorities or weights for each of these characteristics. Statistical methods, such as the design of experiments or factorial design (see Chapter 6 in [3]), then can be used to determine the sensor that is best. Ultimately, the final choice of sensor may involve either some or all of the aforementioned characteristics.

The following example illustrates how the design of a sensor can be a process that often involves reconsideration of the design constraints before arriving at the final design.

## Example Problem 1.1

*Statement*: A design engineer intends to scale down a pressure sensor to fit inside an ultra-miniature robotic device. The pressure sensor consists of a circular diaphragm that is instrumented with a strain gage. The diaphragm is deflected by a pressure difference that is sensed by the gage and transduced by a Wheatstone bridge. The diaphragm of the full-scale device has a 1 cm radius, is 1 mm thick, and is made of stainless steel. The designer plans to make the miniature diaphragm out of silicon. The miniature diaphragm is to have a 600 $\mu$m radius, operate over the same pressure difference range, and have the same deflection. The diaphragm deflection, $\delta$, at its center is

$$\delta = \frac{3(1 - \nu^2)r^4 \Delta p}{16Eh},$$

in which $\nu$ is Poisson's ratio, $E$ is Young's modulus, $r$ is the diaphragm radius, $h$ is the diaphragm thickness, and $\Delta p$ is the pressure difference. Determine the required diaphragm thickness to meet these criteria and comment on the feasibility of the new design.

*Solution*: Assuming that $\Delta p$ remains the same, the new thickness is

$$h_n = h_o \left[ \frac{(1 - \nu_n^2)r_n^4 E_o}{(1 - \nu_o^2)r_o^4 E_n} \right].$$

The properties for stainless steel are $\nu_o = 0.29$ and $E_o = 203$ GPa. Those for silicon are $\nu_n = 0.25$ and $E_n = 190$ GPa. Substitution of these and the aforementioned values into the expression yields $h_n = 1.41 \times 10^{-8}$ m = 14 nm. This thickness is too small to be practical. An increase in $h_n$ by a factor of 10 will increase the $\Delta p$ range likewise. Recall that this design required a similar deflection. A new design would be feasible if the required deflection for the same transducer output could be reduced by a factor of 1000, such as by the use of a piezoresistor on the surface of the diaphragm. This would increase $h_n$ to 14 $\mu$m, which is reasonable using current micro-fabrication techniques. Almost all designs are based upon many factors, which usually require compromises to be made.

## 1.5   Scaling Considerations

Sensors have evolved considerably since the beginning of scientific instruments. Marked changes have occurred in the past 300 years. The temperature sensor serves as a good example. Daniel Gabriel Fahrenheit (1686–1736) produced the first mercury-in-glass thermometer in 1714 with a calibrated scale based upon the freezing point of a certain ice/salt mixture, the freezing point of water, and body temperature. This device was accurate to within several degrees and was approximately the length scale of 10 cm. In 1821, Thomas Johann Seebeck (1770–1831) found that by joining two dissimilar metals at both ends to form a circuit, with each of the two junctions held at a different temperature, a magnetic field was present around the circuit. This eventually led to the development of the thermocouple. Until very recently, the typical thermocouple circuit consisted of two dissimilar metals joined at each end, with one junction held at a fixed temperature (usually the freezing point of distilled water contained within a thermally insulated flask) and the other at the unknown temperature. A potentiometer was used to measure the mV-level emf. Presently, because of the advance in micro-circuit design, the entire reference temperature junction is replaced by an electronic one and contained with an amplifier and linearizer on one small chip. Such chips are being integrated with other micro-electronics and thermocouples such that they can be located in a remote environment and have the temperature signal transmitted digitally with very low noise to a receiving station. The simple temperature sensor has come a long way since 1700.

Sensor development has advanced rapidly since 1990 because of MEMS (microelectromechanical system) sensor technology [1]. The basic nature of sensors has not changed, although their size and applications have changed [4]. Sensors, however, simply cannot be scaled down in size and still operate effectively. Scaling laws for micro-devices, such as those proposed by W.S.N. Trimmer in 1987, must be followed in their design [5]. As sensor sizes are reduced to millimeter and micrometer dimensions, their sensitivities to physical parameters can change. This is because some effects scale with the sensor's physical dimension. For example, the surface-to-volume ratio of a transducer with a characteristic dimension, $L$, scales as $L^{-1}$. So, surface area-active micro-sensors become more advantageous to use as their size is decreased. On the other hand, the power loss-to-onboard power scales as $L^{-4}$. So, as an actuator that carries its own power supply becomes smaller, power losses dominate, and the actuator becomes ineffective. Further, as sensors are made with smaller and smaller amounts of material, the properties of the material may not be isotropic. A sensor having an output that is related to its property values may be less accurate as its size is reduced. For example, the temperature determined from the change in resistance of a miniature resistive element is related to the coefficients of thermal expansion of the material. If property values change with size reduction, further error will be introduced if macro-scale coefficient values are used.

The scaling of most sensor design variables with length is summarized in Table 1.1. This can be used to examine the scaling of some conventional sensors. Consider the laminar flow element, which is used to determine a liquid flow rate. The element basically consists of many parallel tubes through which the bulk flow is subdivided to achieve laminar flow through each tube. The flow rate, Q, is related to the pressure difference, $\Delta p$, measured between two stations separated by a distance, $L$, as

$$Q = C_o \frac{\pi D^4 \Delta p}{128 \mu L},\qquad(1.1)$$

where $D$ is the internal diameter of the pipe containing the flow tubes, $\mu$ is the absolute viscosity of the fluid, and $C_o$ is the flow coefficient of the element. What happens if this device is reduced in size by a factor of 10 in both length and diameter? According to Equation 1.1, assuming $C_o$ is constant, for the same $Q$, a $\Delta p$ 1000 times greater is required! Likewise, to maintain the same $\Delta p$, $Q$ must be reduced by a factor of 1000. The latter is most likely the case. Thus, a MEMs-scale laminar flow element is limited to operating with flow rates that are much smaller than a conventional laminar flow element.

## Example Problem 1.2

*Statement:* Equation 1.1 is valid for a single tube when $C_o = 1$, where it reduces to the Hagen-Poiseuille law. How does the pressure gradient scale with a reduction in the tube's diameter if the same velocity is maintained?

*Solution:* The velocity, $U$, is the flow rate divided by the tube's cross-sectional area, $U = 4Q/(\pi D^2)$, where $D$ is the tube diameter. Thus, Equation 1.1 can be written $\Delta p/L = 32\mu U D^{-2}$. This implies that the pressure gradient increases by a factor of 100 as the tube diameter is reduced by a factor of 10. Clearly, this presents a problem in sensors using micro-capillaries under these conditions. This situation necessitates the development of other means to move liquids in micro-scale sensors, such as piezoelectric and electrophoretic methods.

Decisions on the choice of a micro-sensor or micro-actuator are not based exclusively on length-scaling arguments. Other factors may be more appropriate. This is illustrated by the following example.

## Example Problem 1.3

*Statement:* Most conventional actuators use electromagnetic forces. Are either electromagnetic or electrostatic actuators better for micro-actuators based upon force-scaling arguments?

*Solution:* Using Table 1.1, the electrostatic force scales as $L^2$ and the electromagnetic force as $L^4$. So, a reduction in $L$ by a factor of 100 leads to a reduction in the electrostatic force by a factor of $1 \times 10^4$ and in the electromagnetic force by a factor of $1 \times 10^8$! If these forces are comparable at the conventional scale, then the electrostatic force is 10 000 times larger than the electromagnetic force at this reduced scale.

The final choice of which type of micro-actuator to use, however, may be based upon other considerations. For example, Madou [6] argues that energy density also could be the factor upon which to scale. Energy densities several orders of magnitude higher can be achieved using electromagnetics instead of electrostatics, primarily because of limitations in electrostatic energy density. This could yield higher forces using electromagnetics instead of electrostatics for comparable micro-volumes.

## 1.6 Uncertainty

An important characteristic of a sensor is the uncertainty associated with its use. Many times it is desirable to estimate the uncertainty before deciding

| Variable | Equivalent | L Dimensions | L Scaling |
|---|---|---|---|
| displacement | distance | $L$ | $L$ |
| strain | length change/length | $\Delta L/L$ | $L^0$ |
| strain rate or shear rate | strain change/time | $L^0 T^{-1}$ | $L^0$ |
| velocity | distance/time | $LT^{-1}$ | $L$ |
| surface | width × length | $L^2$ | $L^2$ |
| volume | width × length × height | $L^3$ | $L^3$ |
| force | mass × acceleration | $L^3 LT^{-2}$ | $L^4$ |
| line force | force/length | $L^3 T^{-2}$ | $L^3$ |
| surface force | force/area | $L^3 L^{-1}T$ | $L^2$ |
| body force | force/volume | $L^3 L^{-2}T$ | $L$ |
| work, energy | force × distance | $L^3 L^2 T^{-2}$ | $L^5$ |
| power | energy/time | $L^3 L^2 T^{-3}$ | $L^5$ |
| power density | power/volume | $L^3 L^{-1} T^{-3}$ | $L^2$ |
| electric current | charge/time | $QT^{-1}$ | $L^0$ |
| electric resistance | resistivity × length/cross-sectional area | $L^{-1}$ | $L^{-1}$ |
| electric field potential | voltage | $V$ | $L^0$ |
| electric field strength | voltage/length | $VL^{-1}$ | $L^{-1}$ |
| electric field energy | permittivity × electric field strength$^2$ | $V^2 L^{-2}$ | $L^{-2}$ |
| resistive power loss | voltage$^2$/resistance | $V^2 L$ | $L$ |
| electric capacitance | permittivity × plate area/plate spacing | $L^2 L^{-1}$ | $L$ |
| electric inductance | voltage/change of current in time | $VT^2 Q^{-1}$ | $L^0$ |
| electric potential energy | capacitance × voltage$^2$ | $LV^2$ | $L$ |
| electrostatic potential energy | capacitance × voltage$^2$ with $V \sim L$ | $LV^2$ | $L$ |
| electrostatic force | electrostatic potential energy change/distance | $L^3 L^{-1}$ | $L^2$ |
| electromagnetic force | electromagnetic potential energy change/distance | $L^5 L^{-1}$ | $L^4$ |
| flow rate | velocity × cross-sectional area | $LL^2 T^{-1}$ | $L^3$ |
| pressure gradient | surface force/area/length | $L^2 L^{-1}$ | $L$ |

**TABLE 1.1**
Variable scaling with length, $L$.

upon a particular sensor. In this process, all contributory errors are considered systematic (see Chapter 7 in [3] for detailed uncertainty analysis). This particular type of uncertainty is known as the design-stage uncertainty, $u_d$, which is analogous to the combined standard uncertainty. Often it is used to choose a sensor that meets the accuracy required for a measurement.

Very seldom does the sensor alone comprise the entire measurement system. Its output often is transduced and conditioned before reading. Thus, the uncertainty of the sensor *per se* is only one uncertainty in many that needs to be considered to arrive at the overall uncertainty in the measurement that is made using the sensor. For example, consider the RTD temperature sensor that was described at the beginning of this chapter. Its output is a resistance that is related functionally to its temperature, a coefficient of thermal expansion, and a reference temperature and resistance. Those variables have uncertainties that contribute to the sensor resistance uncertainty. Further, when that sensor is incorporated into a Wheatstone bridge to transduce the sensor resistance into an output bridge voltage, the uncertainties in the bridge's other three resistors and supply voltage must be considered. When the bridge output is read directly by a voltmeter, the uncertainty of the voltmeter is introduced. All of these uncertainties can be quantified by the methods described in the following. This will provide an overall uncertainty for the RTD measurement system.

Often the sensor is combined with a transducer. This package commonly is referred to as an instrument. Its uncertainty is expressed as a function of the zero-order uncertainty of the instrument, $u_0$, and the instrument uncertainty, $u_I$, as

$$u_d = \sqrt{u_0^2 + u_I^2}, \qquad (1.2)$$

which usually is computed at the 95 % confidence level.

Instruments have resolution, readability, and errors. The **resolution** of an instrument is the smallest *physically indicated* division that the instrument displays or is marked. The zero-order uncertainty of the instrument, $u_0$, is set arbitrarily to be equal to one-half the resolution, based upon 95 % confidence. Equation 1.2 shows that the design-stage uncertainty can never be less than $u_0$, which would occur when $u_0$ is much greater than $u_I$. In other words, even if the instrument is perfect and has no instrument errors, its output must be read with some finite resolution and, therefore, some uncertainty.

The **readability** of an instrument is the closeness with which the scale of the instrument is read by an experimenter. This is a subjective value. Readability does *not* enter into assessing the uncertainty of the instrument.

The instrument uncertainty usually is stated by the manufacturer and results from a number of possible elemental instrument uncertainties, $e_i$.

Examples of $e_i$ are hysteresis, linearity, sensitivity, zero-shift, repeatability, stability, and thermal-drift errors. Thus,

$$u_I = \sqrt{\sum_{i=1}^{N} e_i^2}. \tag{1.3}$$

**Instrument errors** (elemental errors) are identified through calibration. An elemental error is an error that can be associated with a *single* uncertainty source. Usually, it is related to the full-scale output (FSO) of the instrument, which is its maximum output value. The most common instrument errors are the following:

1. Hysteresis:

$$\tilde{e}_H = \left(\frac{e_{H,max}}{FSO}\right) = \left(\frac{|y_{up} - y_{down}|_{max}}{FSO}\right). \tag{1.4}$$

The hysteresis error is related to $e_{H,max}$, which is the greatest deviation between two output values for a given input value that occurs when performing an up-scale, down-scale calibration. This is a single calibration proceeding from the minimum to the maximum input values, then back to the minimum. Hysteresis error usually arises from having a physical change in part of the measurement system upon reversing the system's input. Examples include the mechanical sticking of a moving part of the system and the physical alteration of the environment local to the system, such as a region of recirculating flow called a separation bubble. This region remains attached to an airfoil upon decreasing its angle of attack from the region of stall.

2. Linearity:

$$\tilde{e}_L = \left(\frac{e_{L,max}}{FSO}\right) = \left(\frac{|y - y_L|_{max}}{FSO}\right). \tag{1.5}$$

Linearity error is a measure of how linear is the best fit of the instrument's calibration data. It is defined in terms of its maximum deviation distance, $|y - y_L|_{max}$.

3. Sensitivity:

$$\tilde{e}_K = \left(\frac{e_{K,max}}{FSO}\right) = \left(\frac{|y - y_{nom}|_{max}}{FSO}\right). \tag{1.6}$$

Sensitivity error is characterized by the greatest change in the slope (static sensitivity) of the calibration fit.

4. Zero-shift:

$$\tilde{e}_Z = \left(\frac{e_{Z,max}}{FSO}\right) = \left(\frac{|y_{shift} - y_{nom}|_{max}}{FSO}\right). \tag{1.7}$$

Zero-shift error refers to the greatest possible shift that can occur in the intercept of the calibration fit.

5. Repeatability:

$$\tilde{e}_R = \left( \frac{2S_x}{FSO} \right). \tag{1.8}$$

Repeatability error is related to the precision of the calibration. This is determined by repeating the calibration many times for the same input values. The quantity $2S_x$ represents the precision interval of the data for a particular value of $x$.

6. Stability:

$$\tilde{e}_S = \left( \frac{e_{S,max} \cdot \Delta t}{FSO} \right). \tag{1.9}$$

Stability error is related to $e_{S,max}$, which is the greatest deviation in the output value for a fixed input value that could occur during operation. This deviation is expressed in units of $FSO/\Delta t$, with $\Delta t$ denoting the time since instrument purchase or calibration. Stability error is a measure of how much the output for the same input can drift over time since calibration.

7. Thermal-drift:

$$\tilde{e}_T = \left( \frac{e_{T,max}}{FSO} \right). \tag{1.10}$$

Thermal-drift error is characterized by the greatest deviation in the output value for a fixed input value, $e_{T,max}$, that could occur during operation because of variations in environmental temperature. Stability and thermal-drift errors are similar in behavior to the zero-shift error.

The instrument uncertainty, $u_I$, combines all the known instrument errors,

$$u_I = \sqrt{\sum e_i^2} = FSO \cdot \sqrt{\tilde{e}_H^2 + \tilde{e}_L^2 + \tilde{e}_K^2 + \tilde{e}_Z^2 + \tilde{e}_R^2 + \tilde{e}_S^2 + \tilde{e}_T^2 + \tilde{e}_{other}^2}, \tag{1.11}$$

where $\tilde{e}_{other}$ denotes any other instrument errors. All $\tilde{e}_i$'s expressed in Equation 1.11 are dimensionless.

How are these elemental errors actually assessed? Typically, hysteresis and linearity errors are determined by performing a *single* up-scale, down-scale calibration. The results of this type of calibration are displayed in the left graph of Figure 1.2. In that graph, the up-scale results are plotted as open circles and the down-scale results as solid circles. The dotted lines are linear interpolations between the data. Hysteresis is evident in this example by down-scale output values that are higher than their up-scale counterparts. The best-fit curve of the data is indicated by a solid line. Both the hysteresis and linearity errors are assessed with respect to the best-fit curve.

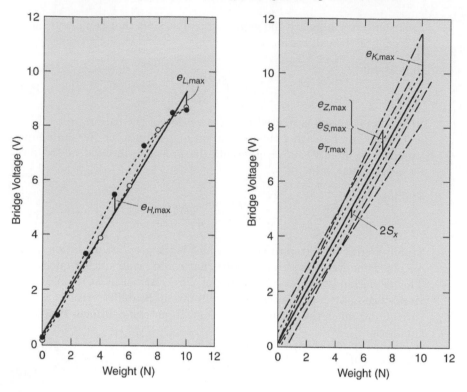

**FIGURE 1.2**
Elemental errors ascertained by calibration.

Sensitivity, repeatability, zero-shift, stability, and thermal-drift errors are ascertained by performing a *series* of calibrations and then determining each particular error by comparisons between the calibrations. The results of a series of calibrations are shown in the right graph of Figure 1.2. The solid curve represents the best fit of the data from *all* the calibrations. The dotted curves indicate the limits within which a calibration is repeatable with 95 % confidence. The repeatability error is determined from the difference between either dotted curve and the best-fit curve. The dash-dot curves identify the calibration curves that have the maximum and minimum slopes. The sensitivity error is assessed in terms of the greatest difference between minimum or maximum sensitivity curve and the best-fit curve. The dashed curves denote shifts that can occur in the calibration because of zero-shift, stability, and thermal-drift errors. Each error can have a different value. It is determined from the greatest difference between calibration data and its best-fit curve.

The following example illustrates the effects of instrument errors on measurement uncertainty.

# Example Problem 1.4

*Statement:* A pressure transducer (a combination of a pressure sensor and a Wheatstone bridge) is connected to a digital panel meter. The panel meter converts the pressure transducer's output in volts back to pressure in psi. The manufacturer provides the following information about the panel meter:

| | |
|---|---|
| Resolution: | 0.1 psi |
| Repeatability: | 0.1 psi |
| Linearity: | within 0.1 % of reading |
| Drift: | less than 0.1 psi/6 months within the 32 °F to 90 °F range |

The only information given about the pressure transducer is that it has "an accuracy of within 0.5 % of its reading."

Estimate the combined standard uncertainty in a measured pressure at a nominal value of 100 psi at 70 °F. Assume that the transducer's response is linear with an output of 1 V for every psi of input.

*Solution:* The uncertainty in the measured pressure, $(u_d)_{mp}$, is the combination of the uncertainties of the transducer, $(u_d)_t$, and the panel meter, $(u_d)_{pm}$. This can be expressed as

$$(u_d)_{mp} = \sqrt{[(u_d)_t]^2 + [(u_d)_{pm}]^2}.$$

For the transducer,

$$(u_d)_t = \sqrt{u_{I_t}^2 + u_{o_t}^2} = u_{I_t} = 0.005 \times 100 \text{ psi} = 0.50 \text{ psi}.$$

For the panel meter,

$$(u_d)_{pm} = \sqrt{u_{I_{pm}}^2 + u_{o_{pm}}^2}.$$

Now,

$$u_{o_{pm}} = 0.5 \text{ resolution} = 0.05 \text{ psi and}$$

$$u_{I_{pm}} = \sqrt{e_1^2 + e_2^2 + e_3^2},$$

where

| | | |
|---|---|---|
| $e_1$ (repeatability) | = | 0.1 psi, |
| $e_2$ (linearity) | = | 0.1 % reading = $0.001 \times 100\text{V}/(1\text{V/psi})$ = 0.1 psi, and |
| $e_3$ (drift) | = | 0.1 psi/6 months $\times$ 6 months = 0.1 psi, |

which implies that

| | | |
|---|---|---|
| $u_{I_{pm}}$ | = | 0.17 psi, |
| $(u_d)_{pm}$ | = | 0.18 psi, and |
| $(u_d)_{mp}$ | = | $\sqrt{0.50^2 + 0.18^2}$ = 0.53 psi. |

Note that most of the combined standard uncertainty comes from the transducer. A more accurate transducer would improve the accuracy of the measurement.

In almost all circumstances, the output of a sensor is related functionally to several variables, such as physical properties and constants, as well as to its input. For example, the change in capacitance of a capacitive pressure transducer (see Chapter 2) is related functionally (see Equation 2.27) to the applied pressure that produces the change in capacitance and to an initial capacitance, an initial pressure, and to five other physical dimensions or properties of the transducer. The uncertainty in the measured pressure can be determined as the uncertainty in a result, where the result is a quantity not measured directly but calculated from measured and known quantities.

The uncertainty in a result, $u_r$, as a function of measurand and other known variable uncertainties, $u_{x_i}$s, is expressed as

$$u_r^2 \simeq u_{x_1}^2 \left(\frac{\partial r}{\partial x_1}\right)^2 + u_{x_2}^2 \left(\frac{\partial r}{\partial x_2}\right)^2 + 2u_{x_1 x_2} \left(\frac{\partial r}{\partial x_1}\right)\left(\frac{\partial r}{\partial x_2}\right) + \dots \ . \qquad (1.12)$$

This equation shows that the uncertainty in the result is a function of the estimated uncertainties (assumed to be variances) $u_{x_1}$ and $u_{x_2}$, and their estimated covariance $u_{x_1 x_2}$. It forms the basis for more detailed uncertainty expressions that can be developed and used to estimate the overall uncertainty in a variable. More formally, $u_r$ is called the **combined standard uncertainty**. In order to determine the combined standard uncertainty, the types of errors that contribute to the uncertainty must be examined first.

General uncertainty analysis is most applicable to experimental situations involving either a single-measurement measurand or a single-measurement result. The uncertainty of a single-measurement measurand is related to its instrument uncertainty, which is determined from calibration, and to the resolution of instrument used to read the measurand value. The uncertainty of a single-measurement result comes directly from the uncertainties of its associated measurands. The expressions for these uncertainties follow directly from Equation 1.12.

For the case of $J$ measurands, the combined standard uncertainty in a result becomes

$$u_r^2 \simeq \sum_{i=1}^{J} (\theta_i)^2 u_{x_i}^2 + 2 \sum_{i=1}^{J-1} \sum_{j=i+1}^{J} (\theta_i)(\theta_j) u_{x_i, x_j}, \qquad (1.13)$$

where

$$u_{x_i, x_j} = \sum_{k=1}^{L} (u_i)_k (u_j)_k, \qquad (1.14)$$

with $L$ being the number of elemental error sources that are common to measurands $x_i$ and $x_j$, and $\theta_i = \partial r / \partial x_i$. $\theta_i$ is the **absolute sensitivity coefficient**. This coefficient should be evaluated at the expected value of $x_i$. Note that the covariances in Equation 1.13 should not be ignored simply for convenience when performing an uncertainty analysis. Variable interdependence should be assessed. This occurs through common factors, such as ambient temperature and pressure for a single instrument used for different measurands.

When the covariances are negligible, Equation 1.13 for $J$ *independent* variables simplifies to

$$u_r^2 \simeq \sum_{i=1}^{J} \left(\theta_i u_{x_i}\right)^2 , \qquad (1.15)$$

where $u_{x_i}$ is the **absolute uncertainty**. From these equations, either the uncertainty in a result, $u_r$, or the **fractional uncertainty** of a result, $u_r/|r|$, can be determined.

---

**Example Problem 1.5**

   *Statement:* The velocity component, $U_\perp$, of a microparticle traveling perpendicularly through the fringes of a laser Doppler velocimeter probe volume is related to the measured Doppler frequency, $f_{diff}$, the beam wavelength, $\lambda$, and the inter-beam angle, $2\kappa$, by the expression

$$U_\perp = \frac{\lambda f_{diff}}{2\sin\kappa}.$$

Determine the fractional uncertainty (in %) of the velocity component, assuming that the fractional uncertainties in the frequency, wavelength, and angle are 2 %, 1 %, and 0.5 %, respectively.

   *Solution:* If $r = (x...z)/(u...w)$, then

$$\frac{u_r}{|r|} = \sqrt{\left(\frac{u_x}{x}\right)^2 + ... + \left(\frac{u_z}{z}\right)^2 + \left(\frac{u_u}{u}\right)^2 + ... + \left(\frac{u_w}{w}\right)^2},$$

which is true for this case. Thus,

$$\frac{u_{U_\perp}}{|U_\perp|} = \sqrt{\left(\frac{u_{f_{diff}}}{f_{diff}}\right)^2 + \left(\frac{u_\lambda}{\lambda}\right)^2 + \left(\frac{u_\kappa}{\kappa}\right)^2}.$$

Substituting the above values gives

$$\frac{u_{U_\perp}}{|U_\perp|} = \sqrt{(0.02)^2 + (0.01)^2 + (0.005)^2} = 0.023.$$

Thus, the fractional uncertainty in the velocity component is 2.3 %.

---

## 1.7   Calibration

Measurement systems and their instruments are used in experiments to obtain measurand values that usually are either steady or varying in time. For both situations, errors arise in the measurand values simply because

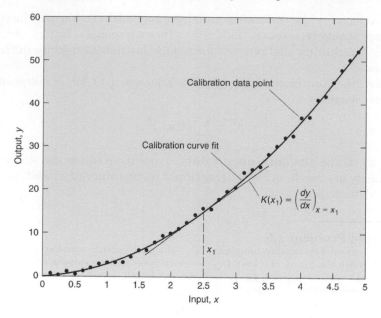

**FIGURE 1.3**
Typical static calibration curve.

the instruments are not perfect; their outputs do not precisely follow their inputs. These errors can be quantified through the process of **calibration**.

In a calibration, a known input value (called the **standard**) is applied to the system and then its output is measured. Calibrations can either be **static** (not a function of time) or **dynamic** (both the **magnitude** and the **frequency** of the input signal can be a function of time). Calibrations can be performed in either **sequential** or **random** steps. In a sequential calibration, the input is increased systematically and then decreased. Usually, this is done by starting at the lowest input value and calibrating at every other input value up to the highest input value. Then the calibration is continued back down to the lowest input value by covering the alternate input values that were skipped during the upscale calibration. This helps to identify any unanticipated variations that could be present during calibration. In a random calibration, the input is changed from one value to another in no particular order.

From a calibration experiment, a **calibration curve** is established. A generic static calibration curve is shown in Figure 1.3. This curve has several characteristics. The **static sensitivity** refers to the slope of the calibration curve at a particular input value, $x_1$. This is denoted by $K$, where $K = K(x_1) = (dy/dx)_{x=x_1}$. Unless the curve is linear, $K$ will not be a constant. More generally, sensitivity refers to the smallest change in a quantity that an instrument can detect, which can be determined knowing the value of $K$

and the smallest indicated output of the instrument. There are two **ranges** of the calibration, the input range, $x_{max} - x_{min}$, and the output range, $y_{max} - y_{min}$.

Calibration accuracy refers to how close the measured value of a calibration is to the **true value**. Typically, this is quantified through the **absolute error**, $e_{abs}$, where

$$e_{abs} = |\text{true value} - \text{indicated value}|. \tag{1.16}$$

The **relative error**, $e_{rel}$, is

$$e_{rel} = e_{abs}/|\text{true value}|. \tag{1.17}$$

The **accuracy** of the calibration, $a_{cal}$, is related to the absolute error by

$$a_{cal} = 1 - e_{rel}. \tag{1.18}$$

Calibration precision refers to how well a particular value is indicated upon repeated but independent applications of a specific input value. An expression for the precision in a measurement and the uncertainties that arise during calibration are presented in Chapter 7 of [3].

## 1.8   Problems

1. Describe the actual sensor in each of the following systems and the basic physical principle or law upon which each works: (a) a tire pressure gage, (b) a toaster, (c) a car mph indicator, (d) a television remote control, and (e) a battery-life indicator on a digital camera.

2. An electrostatic particle levitator operates on the principle of balancing the electrostatic force, $qE$, on a particle with its gravitational force, $mg$. A new, smaller levitator is proposed with both the levitator and particle scaled down geometrically. Assuming the same charge on the particle, determine by using scaling arguments whether or not the smaller levitator will operate the same as the original one.

3. An archaeologist discovers an ancient length-scale bar on which its smallest divisional marks are precisely 10 cm apart. His colleague argues that the civilization that used the bar could measure lengths to an accuracy as small as 0.1 cm by carefully reading in between the precise 10-cm marks. Is the colleague correct?

4. A resistive accelerometer (see Chapter 2) is fabricated with an internal mass of 1 gm and 2-mm-long strain gages, each having a spring constant of 300 N/m. The uncertainties in the mass, strain gage length, and spring constant each are 1 % of their magnitude. Determine the percent uncertainty in the measured acceleration.

5. An ion sensitive electrode (see Chapter 2) is used to measure the $pH$ of a solution in which the hydrogen ion activity is known to within 3 %. Determine the percent uncertainty in the $pH$.

6. A linear spring is extended 0.5 m from its no load position. Its spring constant is 120 N/m. The percent uncertainties in the spring constant and the length measurement are 0.5 % and 1.0 %, respectively. Determine (a) the static sensitivity (in J/m) of the calibration curve of spring *energy* versus extension distance at the 0.5 m extension and (b) the uncertainty in the static sensitivity (in J/m) at that extension.

# Bibliography

[1] Kovacs, G.T.A. 1998. *Micromachined Transducers Sourcebook.* New York: McGraw-Hill.

[2] White, R.M. 1987. A Sensor Classification Scheme. *IEEE Transactions on Ultrasonics, Ferroelectrics and Frequency Control* UFFC-34: 124–126.

[3] Dunn, P.F. 2010. *Measurement and Data Analysis for Engineering and Science.* 2nd ed. Boca Raton: CRC Press: Taylor and Francis Group.

[4] Vetelino, J. and Reghu, A. 2011. *Introduction to Sensors.* New York: CRC Press.

[5] Hsu, T.-R. 2002. *MEMS & Microsystems: Design and Manufacture.* New York: McGraw-Hill.

[6] Madou, M. 1997. *Fundamentals of Microfabrication.* New York: CRC Press.

# 2

# Sensors in Engineering and Science

## CONTENTS

## 2.1   Chapter Overview

Sensors can be understood best by examining the basic physical principles upon which they are designed. In this chapter, some sensors involved in the measurement of length, relative displacement, force, pressure, acceleration, sound pressure, velocity, volumetric and mass flow rates, temperature, heat flux, relative humidity, circular frequency, particle diameter, void fraction, density, density gradient, gas concentration, and pH are presented. The fundamental equations that relate what is sensed to its measurable output are given for each sensor described.

## 2.2   Physical Principles of Sensors

The first step in choosing a sensor is to gain a thorough understanding of the basic physical principle behind its design and operation. The principles of sensors [1] do not change. However, their designs change almost daily. Once the basic principles are understood, then the web sites of most sensor and transducer manufacturers can be consulted. The websites provide current information on their products and performance characteristics. Further, many standard measurement textbooks, for example, [2], [3], [4], [5], and [6], can be consulted to obtain descriptions of innumerable devices based upon these principles. So, if the basic principles are understood, then it is straightforward to identify a sensor suitable for the intended purpose.

The choice of either selecting or designing a sensor starts with identifying the physical variable to be sensed and the physical principle or law associated with that variable. Sensors can be developed for different measurands and can be based upon the same physical principle or law. For example, a thin-wire sensor's resistance inherently changes with strain. This wire can be mounted on various structures and used with a Wheatstone bridge to measure strain, force, pressure, torque, or acceleration. Likewise, sensors can be developed for the same measurand and can be based upon different physical principles or laws. For example, a thin wire's resistance also inherently changes with temperature. This, as well as other sensors, such as a thermocouple, a thermistor, a constant-current anemometer, and an optical pyrometer, can be used to measure temperature.

In the remainder of this chapter, the sensors described are organized with respect to the physical basis by which they sense. Each sensor is listed in Table 2.1 according to its physical principle or method along with its measurand. The measurands, associated variables, and sensor name abbreviations are $T$ (temperature), $U$ (velocity), $q$ (heat flux), $L$ (length), $\delta$ (relative displacement), $F$ (force), $p$ (pressure), $a$ (acceleration), $h$ (distance), $\epsilon_r$ (relative dielectric constant), $A$ (area), $sp$ (sound pressure), $RH$ (relative humidity), $\omega$ (circular frequency), $Q$ (volumetric flow rate), $\dot{m}$ (mass flow rate), $d_p$ (particle diameter), $\alpha$ (void fraction), $\rho$ (density), $d\rho/dx$ (density gradient), $[C]$ (gas concentration), $\epsilon$ (strain), $pH$ (-$\log_{10}$ [H$^+$]), DI (displacement indicator), and LVDT (linear variable differential transformer). There are an uncountable number of other sensors available today. Often, full descriptions and performance characteristics of sensors offered by manufacturers are available on the Internet.

## 2.3 Electric

Sensors in this category are based upon a change in either resistance, capacitance, or inductance that results from applying a physical stimulus. Their output, in some instances, can be read directly, such as the resistance of a **resistance temperature detector** or **resistive thermal device** (RTD) using an ohmmeter. Typically, however, they are configured as part of an electrical circuit that requires a power supply to produce either a voltage or current output.

| Basis | Sensor | Measurand |
|---|---|---|
| electric: resistive | resistance temperature device | $T$ |
| " | hot-wire or hot-film probe | $T, U, q$ |
| " | thermistor | $T$ |
| " | strain gage | $\delta$ |
| " | resistance DI | $\delta$ |
| " | resistance force transducer | $F$ |
| " | resistance pressure transducer | $p$ |
| " | resistance accelerometer | $a$ |
| electric: capacitive: $\Delta h$ | capacitive DI | $L$ |
| " | capacitive level indicator | $L$ |
| " | capacitive pressure transducer | $p$ |
| " | capacitive accelerometer | $a$ |
| " | capacitive microphone | $sp$ |
| electric: capacitive: $\Delta \epsilon_r$ | capacitive DI | $\delta$ |
| " | relative humidity indicator | $RH$ |
| electric: capacitive: $\Delta A$ | capacitive DI | $\delta$ |
| electric: inductive | variable reluctance DI | $\delta$ |
| " | LVDT | $\delta$ |
| " | reluctance tachometer | $\omega$ |
| piezoelectric | piezoresistive pressure transducer | $p$ |
| " | piezoresistive accelerometer | $a$ |
| fluid mechanic | Pitot-static tube | $p$ |
| " | venturi | $Q$ |
| " | flow nozzle | $Q$ |
| " | orifice plate | $Q$ |
| " | laminar flow element | $Q$ |
| " | rotameter | $Q$ |
| " | vortex-shedding flowmeter | $Q$ |
| " | Coriolis flowmeter | $\dot{m}$ |
| " | turbine flowmeter | $\dot{m}$ |
| optic | laser Doppler velocimeter | $U$ |
| " | phase Doppler anemometer | $U, d_p$ |
| " | particle image velocimeter | $U$ |
| " | particle laser light sheet | $U$ |
| " | transmitted and reflected light | $\alpha$ |
| " | interferometer | $\rho$ |
| " | Schlieren | $d\rho/dx$ |
| " | nondispersive infrared detector | $[C]$ |
| photoelastic | plane polarizer | $\epsilon$ |
| " | moiré method | $\epsilon$ |
| thermoelectric | thermocouple | $T$ |
| electrochemical | Taguchi sensor | $[C]$ |
| " | ion sensitive electrode | $pH$ |
| " | ChemFET sensor | $[C]$ |

**TABLE 2.1**

Sensors described in this chapter and their measurands.

## 2.3.1   Resistive

A sensor based upon the principle that a change in resistance can be produced by a change in a physical variable is, perhaps, the most common type of sensor. A resistance sensor can be used to measure displacement, strain, force, pressure, acceleration, flow velocity, temperature, heat or light flux, and gas concentration.

The resistance of a conductor or semiconductor changes measurably with temperature. One simple sensor of this type is a metal wire or conducting strip. Its resistance is related to temperature by

$$R = R_o[1 + \alpha(T - T_o) + \beta(T - T_o)^2 + \gamma(T - T_o)^3 + ...], \qquad (2.1)$$

where $\alpha, \beta$, and $\gamma$ are coefficients of thermal expansion, and $R_o$ is the resistance at the reference temperature $T_o$. In many situations having a temperature range of approximately 50 K or less, the higher-order terms are negligible and Equation 2.1 reduces to the linear expression

$$R \simeq R_o[1 + \alpha(T - T_o)]. \qquad (2.2)$$

The resistance of a semiconductor as a function of temperature is described implicitly by the Steinhart-Hart equation, where

$$\frac{1}{T} = A + B \ln R + C(\ln R)^3, \qquad (2.3)$$

in which $A$, $B$, and $C$ are the Steinhart-Hart coefficients. Typically, these coefficients are determined for a material by measuring its resistances at three different temperatures, then solving the resulting three equations.

### Resistance Temperature Detector

One sensor based upon this principle includes the **RTD** for temperature measurement. The sensing element of the RTD can be a wire wound around an insulating support, a thin wire, a thin strip of metal, or a thin insulating strip with a deposited conducting film. The metals usually are platinum, copper, or nickel, having coefficients of thermal expansion of 0.0039/°C, 0.0038/°C, and 0.0067/°C, respectively.

---

**Example Problem 2.1**

*Statement:* A platinum RTD is used to determine the liquid temperature of a yeast-proofing mixture. The measured resistance of the RTD is 25.0 Ω at 0 °C and 29.8 Ω at the proofing temperature. What is the proofing temperature (in K)?

*Solution:* Equation 2.2 can be used for this situation. This equation can be rearranged to become

$$T = T_o + \left[\frac{R - R_o}{\alpha R_o}\right].$$

Thus, $T$ (°C) $= 0 + (29.8\text{-}25.0)/[(0.0039)(25.0)] = 49.2$ °C $= 322.4$ K.

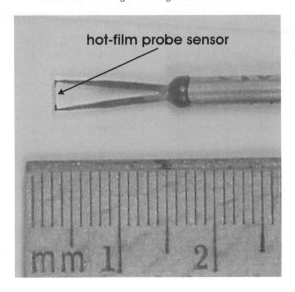

**FIGURE 2.1**
The quartz-coated hot-film probe.

### Hot-Wire or Hot-Film Anemometer

Another sensor based upon resistance change with temperature is the **hot-wire** or **hot-film probe**, which is part of an anemometry system. This can be used to determine temperature, velocity, and heat flux. The sensor is either a wire (as small as ~25 $\mu$m in diameter) or a very small rod (as small as ~250 $\mu$m in diameter) covered with a thin conducting film and then an electrical insulator (see Figure 2.1 and [7]). The former is termed a **hot-wire sensor** and the latter a **hot-film sensor**. The coefficients of thermal expansion and several values of the resistance as a function of temperature usually are provided by the manufacturer.

*Constant-Temperature Mode*: The **constant-temperature mode** is used to measure the local velocity in a fluid flow. The sensor is connected to a Wheatstone bridge feedback amplifier circuit that is used to maintain the sensor at a constant resistance and, hence, at a constant temperature above the fluid's temperature. As the sensor is exposed to different velocities, the power required to maintain the wire at the constant temperature changes because of the changing heat transfer to the environment. This can be expressed by applying the conservation of energy to the sensor and its surroundings, equating the power supplied to the sensor to its rate of heat transfer to the surrounding fluid by convection. This becomes

$$E_s^2/R_s = h_{sf}A_s(T_s - T_f), \tag{2.4}$$

in which $E_s$ is the potential difference (voltage) across the sensor, $R_s$ is the sensor resistance, $h_{sf}$ is the convective heat transfer coefficient for the sensor and the fluid, $A_s$ is the surface area of the sensor over which heat is transferred to the fluid, $T_s$ is the sensor temperature, and $T_f$ is the fluid temperature.

The potential difference across the sensor is related to the output voltage of the bridge circuit, $E_o$, by

$$E_s = E_o R_s / (R_s + R_o),  \qquad (2.5)$$

where $R_o$ denotes the resistance that is in series with $R_s$ in the bridge. The heat transfer coefficient for a small wire or rod is

$$h_{sf} = C_0 + C_1 \sqrt{\rho U},  \qquad (2.6)$$

in which $C_0$ and $C_1$ are constants that depend upon the ambient temperature, $\rho$ the density of the fluid, and $U$ the local fluid velocity near the sensor. Substitution of Equations 2.5 and 2.6 into Equation 2.4 yields

$$E_o^2 = A + B \sqrt{\rho U},  \qquad (2.7)$$

in which $A = C_0 (R_o + R_s)^2 (T_s - T_f)/R_s$ and $B = (C_1/C_0)A$. $A$ and $B$ are both constant because the sensor temperature and its resistance is held constant. Equation 2.7 is known as King's law, which was derived by King in the early 1900s. It shows that the bridge voltage is proportional to the square root of the velocity. Additional example problems involving the constant-temperature mode are presented in [8].

*Constant-Current Mode*: The **constant-current mode** is used to determine temperature, velocity, and rate of heat transfer. This employs a different circuit having a battery in series with a variable resistor (a high-impedance current source) and the sensor, thereby supplying a constant current to the sensor. The voltage across the sensor is amplified and measured. As the fluid velocity changes, the rate of heat transfer changes, causing a change in the sensor temperature and resistance. Using values of the known constant current, $I_c$, and the measured voltage across the sensor, the sensor resistance becomes $E_s/I_c$. The sensor temperature can then be determined using Equation 2.4. The rate of heat transfer equals the power supplied to the sensor, $I_c^2 R_s$. The velocity can be determined using Equations 2.4 and 2.6 and known constants. Although this mode appears more versatile than the constant-temperature mode, the latter has a much higher frequency response when measuring velocity. Hence, the constant-temperature mode is used more often.

---

## Example Problem 2.2

*Statement:* A hot-film anemometry system is used in the constant-current mode to determine the convective heat transfer coefficient of the hot-film sensor to liquid mercury. Develop a method to compute the convective heat transfer coefficient in terms

of known and measured variables. Then determine its value for the measured values
of $E_s = 22$ V, $I_c = 2.44$ A, and a fluid temperature, $T_f$, of 300 K. Assume that the
reference values for the hot-film probe are the sensor length, $L_s = 206$ mm, sensor
diameter, $d_s = 2$ mm, $R_o = 9$ Ω, $T_o = 280$ K, and $\alpha = 0.003/$K.

*Solution:* The power supplied to the sensor equals the rate of heat transfer from
the sensor to the mercury. Thus,

$$I_c^2 R_s = h_{sf} A_s (T_s - T_f).$$

Rearranging this equation gives

$$h_{sf} = I_c^2 R_s / [A_s (T_s - T_f)].$$

Because the current, $I_c$, and voltage across the sensor, $E_s$, are measured, the sensor
resistance, $R_s$, is known and equals $E_s/I_c$. The sensor temperature can be determined
using Equation 2.2 and the reference conditions of the hot-film sensor. Substitution of
the obtained values into the above expression for $h_{sf}$ will yield the desired value.

For the given values, $R_s = E_s/I_c = 8.98$ Ω. This yields $T_s = T_o + (R_s - R_o)/(\alpha R_o)$
$= 320.8$ K. Thus, $h_{sf} = I_c^2 R_s / [A_s (T_s - T_f)] = 1990$ W/(m·K).

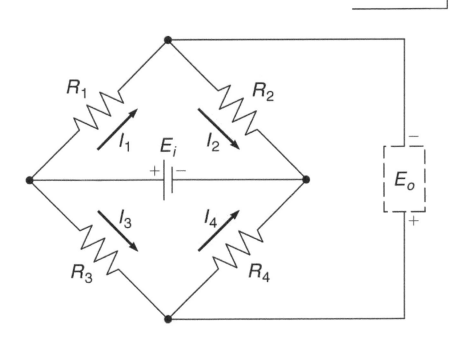

**FIGURE 2.2**
The Wheatstone bridge configuration.

*Bridge-Balance Mode*: A third operational mode is the **bridge-balance
mode** or **null mode**. It is used to measure fluid temperature. Referring to
Figure 2.2, a variable resistor (corresponding to $R_4$) is used in the leg of the
Wheatstone bridge that is opposite the sensor (corresponding to $R_1$). When
the bridge is balanced, the product of the variable and sensor resistances,
$R_1 R_4$, equals the product of the two other resistances in the bridge, $R_2 R_3$,

which is constant. Thus, by measuring value of the variable resistance that is required to balance the bridge, the sensor resistance, and, hence, its temperature is determined. For this situation, the sensor is in thermal equilibrium with the fluid. So, the sensor temperature is the fluid temperature. Several examples of Wheatstone bridge methods are covered in Chapter 2 of [8].

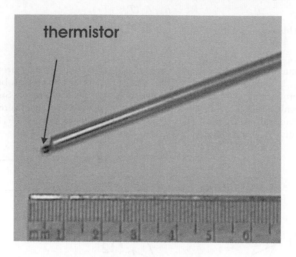

**FIGURE 2.3**
A thermistor (~1 mm diameter) with supporting tubing.

**Thermistor**
A **thermistor** is another sensor that is used to measure temperature. It consists of two wires that are connected across a small piece of semiconductor material (see Figure 2.3).

Thermistors can have either a decreasing or an increasing resistance with temperature. The former is called a negative temperature coefficient (NTC) thermistor and the latter a positive temperature coefficient (PTC) thermistor. The resistance-temperature behavior of a thermistor is described by Equation 2.3. That for the more commonly used NTC can be expressed in the reduced form

$$R = R_o \exp\left[\beta(\frac{1}{T} - \frac{1}{T_o})\right], \tag{2.8}$$

in which $A = (1/T_o) - (\ln R_o)/\beta$, $B = 1/\beta$, and $C = 0$ have been substituted into Equation 2.3. Typical magnitudes of $A$, $B$, and $C$ are $1 \times 10^{-3}$, $2 \times 10^{-4}$, and $9 \times 10^{-8}$, respectively, in units of K$^{-1}$.

The thermistor usually gives better resolution over a small temperature range because of its resistance's exponential change with temperature, whereas the RTD can cover a wider temperature range. For both sensors, a transducer such as a Wheatstone bridge circuit typically is used to convert resistance to voltage. In this bridge circuit, either sensor would serve as $R_1$ as shown in Figure 2.2.

### Strain Gage

The resistance of a conductor also changes measurably with strain. When a fine wire of length $L$ is stretched, its length increases by $\Delta L$, yielding a longitudinal strain of $\epsilon_L \equiv \Delta L/L$. This produces a change in resistance. Its width decreases by $\Delta d/d$, where $d$ is the wire diameter. This defines the transverse strain $\epsilon_T \equiv \Delta d/d$. Poisson's ratio, $\nu$, is defined as the negative of the ratio of transverse to longitudinal *local* strains, $-\epsilon_T/\epsilon_L$. The negative sign compensates for the *decrease* in transverse strain that accompanies an *increase* in longitudinal strain, thereby yielding positive values for $\nu$. Poisson's ratio is a material property that couples these strains.

For a wire, the resistance $R$ can be written as

$$R = \rho\frac{L}{A}, \tag{2.9}$$

in which $\rho$ is the resistivity, $L$ the length, and $A$ the cross-sectional area. Taking the total derivative of Equation 2.9 yields

$$dR = \frac{\rho}{A}dL + \frac{L}{A}d\rho - \frac{\rho L}{A^2}dA. \tag{2.10}$$

Equation 2.10 can be divided by Equation 2.9 to give the relative change in resistance,

$$\frac{dR}{R} = (1 + 2\nu)\epsilon_L + \frac{d\rho}{\rho}. \tag{2.11}$$

Equation 2.11 shows that the relative resistance change in a wire depends on the strain of the wire and the resistivity change.

The **strain gage** is the most frequently used resistive sensor. A typical strain gage is shown in Figure 2.4. The gage consists of a very fine, etched wire of length $L$ that winds back and forth over a flat, insulating sensing area. For the strain gage shown, there are 12 wire segments, yielding a total wire length of $\sim$12 cm.

A **local gage factor**, $G_l$, can be defined as the ratio of the relative resistance change to the relative length change,

$$G_l = \frac{dR/R}{dL/L}. \tag{2.12}$$

This expression relates *differential* changes in resistance and length and describes a factor that is valid only over a very small range of strain.

An **engineering gage factor**, $G_e$, can be defined as

$$G_e = \frac{\Delta R/R}{\Delta L/L}. \tag{2.13}$$

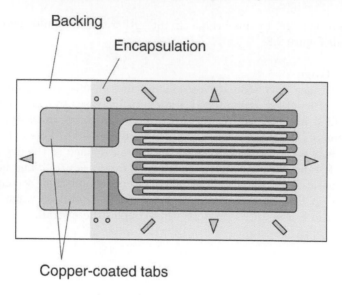

Copper-coated tabs

**FIGURE 2.4**
A strain gage with a typical sensing area of 5 mm × 10 mm.

This expression is based on small, finite changes in resistance and length. This gage factor is the slope based on the total resistance change throughout the region of strain investigated. The local gage factor is the instantaneous slope of a plot of $\Delta R/R$ versus $\Delta L/L$. Because it is very difficult to measure local changes in length and resistance, the engineering gage factor typically is used more frequently. Equation 2.11 can be rewritten in terms of the engineering gage factor as

$$G_e = 1 + 2\nu + \left[ \frac{\Delta\rho}{\rho} \cdot \frac{1}{\epsilon_L} \right]. \tag{2.14}$$

For most metals, $\nu \approx 0.3$. The last term in brackets represents the strain-induced changes in the resistivity, which is a piezoresistive effect (see Section 2.4). This term is constant for typical strain gages and equals approximately 0.4. Thus, the value of the engineering gage factor is approximately 2 for most metallic strain gages.

**Resistive Displacement Indicator**

An alternative expression for the relative change in resistance can be derived using statistical mechanics, where

$$\frac{dR}{R} = 2\epsilon_L + \frac{dv_0}{v_0} - \frac{d\lambda}{\lambda} - \frac{dN_0}{N_0}. \tag{2.15}$$

Here $v_0$ is the average number of electrons in the material in motion between ions, $\lambda$ is the average distance traveled by an electron between collisions, and $N_0$ is the total number of conducting electrons. Equation 2.15 implies that the differential resistance change and, thus, the gage factor, is independent of the material properties of the conductor. This also implies that the change in resistance only will be proportional to the strain when the sum of the differential changes on the right hand side of Equation 2.15 is either zero or directly proportional to the strain. Fortunately, most strain gage materials have this behavior. So, when a strain gage is used in a circuit such as a Wheatstone bridge, strain can be converted into a voltage. This system can be used as a **resistive displacement indicator.**

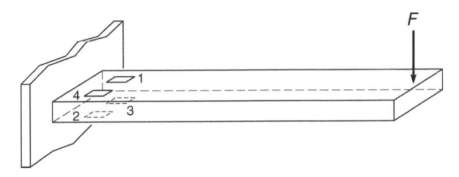

**FIGURE 2.5**
Cantilever beam with four strain gages.

### Resistive Force Transducer

Strain gages also can be mounted on a number of different flexures to yield various types of sensor systems. One example is four strain gages mounted on a beam to determine its deflection, as described in detail in Chapter 2 of [8] and shown in Figure 2.5. This is a **resistive force transducer**, although it actually contains sensors (four strain gages) and a transducer (the Wheatstone bridge).

Each of the strain gages serves as resistances in the four legs of a Wheatstone bridge (refer to Figure 2.2). As force is applied to the top edge of the beam, it deflects downward, producing a strain on the top two gages and a compression on the bottom two gages. Strain is converted into an increase in gage resistance and compression into a decrease. If each of the four gages has the same resistance, $R$, when no force is applied, then the gage resistance will increase by $\delta R$ for the top gages and decrease by $\delta R$ for the bottom two gages. Using this deflection method, the Wheatstone bridge equation

$$E_o = E_i \left[ \frac{R_1}{R_1 + R_2} - \frac{R_3}{R_3 + R_4} \right] \tag{2.16}$$

reduces to

$$E_o = E_i(\delta R/R), \tag{2.17}$$

in which $E_i$ is the input voltage to the bridge. For a cantilever beam supported at its end with a force applied at the center of its other end, the top-side strain and bottom-side compression are proportional to the applied force, $F$. If the strain gages are aligned with this axis of strain, then $\delta R \sim \epsilon_L$. Hence, $E_o \sim \delta R \sim \epsilon_L \sim F$. Thus, $E_o = kF$, where $k$ is the static sensitivity of the system calibration's linear fit. Further, with this strain gage configuration, variational temperature and torsional effects are compensated for automatically.

### Resistive Pressure Transducer

Another example involves one or more strain gages mounted on the surface of a diaphragm that separates two chambers exposed to different pressures. As the diaphragm is deflected because of a pressure difference between the two chambers, a strain is produced. For a circular diaphragm that is supported along its circumference, the steady-state pressure, $\Delta P$, is related to the diaphragm deflection at its center, $\delta y_c$, as

$$\Delta P = \left[\frac{16Et^2}{3r^4(1-\nu^2)}\right] y_c[1 + (y_c/4t)^2], \tag{2.18}$$

in which $E$ is Young's modulus, $t$ the thickness, $r$ the radius, and $\nu$ Poisson's ratio. If the diaphragm is fabricated such that the ratio $y_c/4t$ becomes negligible with respect to unity, then $\Delta P$ is linearly proportional to $y_c$. This, however, reduces the frequency response of the diaphragm, which may be important in measuring non-steady pressures. If a strain gage is mounted on the surface of the diaphragm (sometimes, a circular strain gage is used), then its change in resistance from the zero-deflection case will be proportional to $y_c$. The resultant change in resistance usually is converted into a voltage using a Wheatstone bridge. This is called a **resistive pressure transducer**, although it actually contains both a sensor (the strain gage) and a transducer (the Wheatstone bridge).

### Resistive Accelerometer

A **resistive accelerometer** uses a strain gage flexure arrangement. An accelerometer in the 1970s typically contained a small mass that was moved against a spring as the device containing them was accelerated. The displacement of the mass was calibrated against a known force. This information then was used to determine the acceleration from the displacement using Newton's second law. Today, strain-gage accelerometers are common, especially because of their reduced size. In these accelerometers, strain gages have replaced springs.

A typical strain-gage accelerometer uses four similar strain gages, as shown in Figure 2.6. Two very fine wires, which serve as strain gages and

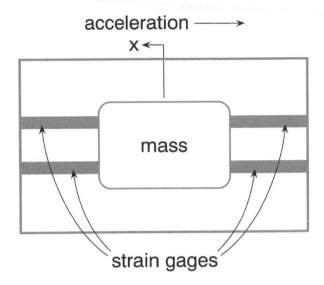

**FIGURE 2.6**
An accelerometer with four strain gages.

springs, are located on two opposing sides of a small mass, $m$, along the axis of acceleration. The mass and wires are contained inside a housing that is filled with a fluid to provide moderate damping. The system essentially behaves as a spring-mass system. (See Chapter 4 of [8] for the governing equations and response characteristics of such systems.) The displacement, $\Delta L$, of the mass, $m$, due to an acceleration, $a$, is

$$\Delta L = (m/k)a, \tag{2.19}$$

in which $k$ is the spring constant (effective stiffness) of the wires. The strain, $\epsilon$, then becomes

$$\epsilon \equiv \Delta L/L = (m/kL)a. \tag{2.20}$$

Thus, the strain is linearly proportional to the acceleration. If the four gages are connected to a Wheatstone bridge using the deflection method, the output voltage of the system will be linearly proportional to the acceleration.

## 2.3.2 Capacitive

Other types of sensors produce a change in capacitance with the change in a physical variable. A capacitive sensor can be used to measure level, displacement, strain, force, pressure, acceleration, and humidity.

A **capacitive sensor** consists of two small conducting plates, each of area, $A$, separated by a distance, $h$, with a dielectric material in between the two plates. The capacitance between the two plates is

$$C = \epsilon_o \epsilon A / h, \tag{2.21}$$

in which $\epsilon_o$ is the permittivity of free space and $\epsilon$ the relative permittivity. Thus, a change in either $\epsilon$, $A$, or $h$ will produce a change in capacitance. Differentiating Equation 2.21 with respect to $C$ gives

$$\frac{dC}{C} = \frac{d\epsilon}{\epsilon} + \frac{dA}{A} - \frac{dh}{h}. \tag{2.22}$$

Each of these changes has been exploited in developing different capacitive-based sensors.

***Variable-Spacing***: If $\epsilon$ and $A$ are constant, then

$$\frac{\Delta C}{C_o} = \frac{1}{1 \pm (h_o / \Delta h)}, \tag{2.23}$$

in which $C_o$ and $h_o$ denote the initial, unchanged state, and the $\pm$ indicates either an increase or a decrease in $h$ from $h_o$, respectively. When $\Delta h << h_o$, then Equation 2.23 reduces to

$$\frac{\Delta C}{C} = -\frac{\Delta h}{h_o}. \tag{2.24}$$

For this situation, the change in capacitance is linearly proportional and opposite in sign to the change in spacing.

### Capacitive Displacement Indicator

A **capacitive displacement indicator** system used to sense small displacements can be made using a Wheatstone bridge. Referring to Figure 2.2, this is done by having a variable-spacing capacitor (the sensor) in one leg of a Wheatstone bridge ($C_4$ that replaces $R_4$) and using an ac power supply for $E_i$. Another capacitor with fixed spacing and the same initial capacitance as the sensor is placed in the other leg of the bridge ($C_3$ replaces $R_3$) that is in series with the sensor between the power supply. Two resistors having the same resistance are placed in the other two legs of the bridge ($R_1$ and $R_2$).

For a change in capacitance, $\Delta C$, in the sensor, the Wheatstone bridge equation becomes

$$E_o = E_i \left[ \frac{\Delta C}{4C + 2\Delta C} \right]. \tag{2.25}$$

If the change in capacitance is negligibly small with respect to the initial capacitance, then Equation 2.25 reduces to

$$E_o = E_i \left[ \frac{\Delta C}{4C} \right].$$ (2.26)

Thus, the output voltage of the bridge will be proportional to the *change* in capacitance and, hence, to a small displacement.

Another displacement indicator system that is linear over a wider operating range of displacements can be made by including a third plate that can be moved between the capacitor's two fixed plates. This is a **differential capacitive displacement indicator** [3]. Also, a **capacitive level indicator** can be constructed using two concentric conducting cylinders. The gap between the cylinders contains a non-conducting liquid of unknown level, and the remaining space is air. The total capacitance, which is the sum of the liquid and air capacitances, can be related to the liquid level. When this sensor is used as part of a Wheatstone bridge in the deflection mode, the voltage output can be related to the fluid level [3].

### Capacitive Pressure Transducer

The variable-spacing capacitor also can be used as a sensor in a system to measure pressure. For this capacitor, the dielectric is air and one plate is held fixed. As the other plate moves because of the forces acting on it, the capacitance of the sensor changes. Applying Equation 2.18, which describes the plate deflection, for small displacements, the change (decrease) in capacitance, $\Delta C$, is related to the pressure, $P$, producing the change by

$$\frac{\Delta C}{C} = \frac{3a^4(1 - \nu^2)}{16hEt^3}(P - P_o).$$ (2.27)

Thus, the change in capacitance is proportional to $P - P_o$, where $P - P_o$ is the difference between the applied pressure, $P$, and the reference pressure measured at zero plate deflection, $P_o$. When used in a capacitive Wheatstone bridge circuit similar to that described for the capacitive displacement indicator, the pressure change is converted into a voltage. This system forms a **capacitive pressure transducer**.

### Capacitive Accelerometer

A central plate fixed to a small mass can be used instead of air between the two capacitor plates. As the mass and its attached central plate are accelerated, the change in capacitance with respect to time is sensed. The change in capacitance is converted into a voltage that is proportional to the acceleration. This is done using an ac-powered Wheatstone bridge in a manner similar to that done for the change in resistance of a strain-gage accelerometer. This system constitutes a **capacitive accelerometer**.

### Capacitor Microphone

Sound pressure can be measured using a **capacitor microphone**. The sensor effectively serves as a capacitor formed by a thin ($\sim$0.001 in.), stretched diaphragm, a small air gap (~0.001 in.), and a receiving plate. The plate is connected through a circuit capacitor to an amplifier and through a circuit resistor to a dc-voltage source that is biased to several hundred volts. This bias voltage establishes a fixed charge on the sensor's diaphragm and receiving plate. The source cathode is connected to the amplifier's other input and its anode to the circuit resistor and the sensor housing.

The impingement of sound pressure waves on the sensor causes small displacements of the diaphragm and, hence, relatively small variations in the sensor's capacitance, as described by Equation 2.27. Because the charge is fixed, any change in capacitance results in a change in voltage. This voltage can be related directly to the sound pressure using Equation 2.27 and noting that $\Delta V = q/\Delta C$.

There are many other types of sound pressure sensors. Two other frequently used types are the **electret microphone** and the **piezoelectric microphone**. In the former, a ferromagnetic material that contains a permanent charge (the electret), or in the latter, a piezoelectric crystal with face electrodes replaces the capacitor and the voltage source of the capacitor microphone.

*Variable Dielectric*: Capacitive sensors have been made to sense displacement and relative humidity. These sensors exploit the effect on capacitance of varying the permittivity of the dielectric between the capacitor's two conducting plates.

### Capacitive Displacement Indicator

A capacitor can be constructed such that the dielectric between the two plates is solid and can be moved partially outside the plate, with the vacant area occupied by a gas. If $L$ is the length of the plates and the dielectric is moved outside the plates a distance $x$, then the area of the gas between the plates will be $wx$, where $w$ is the depth of the plates and dielectric. The remaining area of the dielectric between the plates will be $w(L - x)$. Both the gas and the dielectric each contribute to the total capacitance, which is the sum of the two capacitances because they electrically are in parallel. Thus,

$$C = \frac{\epsilon_o \epsilon_1 wx}{h} + \frac{\epsilon_o \epsilon_2 w(L - x)}{h}, \tag{2.28}$$

in which $\epsilon_1$ and $\epsilon_2$ are the permittivities of the gas and solid dielectric, respectively. This equation can be rearranged to become

$$C = \frac{\epsilon_o w}{h} \left[ \epsilon_2 L - (\epsilon_2 - \epsilon_1)x \right]. \tag{2.29}$$

Thus, the capacitance is linearly proportional to the displacement, $x$.

### Relative Humidity Sensor

Another type of capacitive sensor is the thin film capacitive **relative humidity sensor**[3]. The dielectric material of this sensor can readily absorb and desorb water molecules, thereby changing its permittivity and, thus, the capacitance of the sensor. The air's relative humidity is proportional to its water molecule concentration. The capacitance of the sensor is related linearly to the relative humidity as

$$C = A + B \cdot RH, \tag{2.30}$$

in which $A$ and $B$ are constants and RH is the % relative humidity.

*Variable Area*: Capacitive sensors can be made to sense displacement by moving one conducting plate laterally with respect to the other plate. This effectively reduces that overlap area that is between the two plates and, hence, the capacitance.

### Capacitive Displacement Indicator

A **capacitive displacement indicator** can be developed in the variable-area principle. If the top plate of area $wL$ is moved laterally a distance $x$, the overlap area between the plates decreases by $wx$, where $w$ is the width of the plate and $L$ is its length. The capacitance then becomes

$$C = \frac{\epsilon_o \epsilon w}{h}(L - x). \tag{2.31}$$

Thus, capacitance is linearly proportional to the displacement and decreases with increasing displacement.

## 2.3.3 Inductive

Some electrical output sensors exploit the change in inductance that occurs with displacement. These are used primarily to measure displacement and strain.

When an ac current is passed through a wire wound as a coil, it generates a magnetic field in and surrounding the coil. The resulting magnetic flux, $\phi_m$, is proportional to the current, $I$. Formally,

$$\phi_m = LI, \tag{2.32}$$

in which $L$ is the inductance. Its unit is the henry (H), which is a weber (Wb) per ampere (A). Equation 2.32 implies that

$$\frac{d\phi_m}{dt} = L\frac{dI}{dt}. \tag{2.33}$$

The time-varying flux induces an electromotive force (emf), E, that opposes the flow of current. This is expressed through Faraday's law of induction as

$$E = -n\frac{d\phi_m}{dt},$$
(2.34)

in which $n$ is the number of turns in the coil. Further, if the length and cross-sectional area of the flux path are $L_f$ and $A_f$, respectively, then the reluctance, R, of the circuit is

$$R = L_f/(\mu_o\mu A_F),$$
(2.35)

in which $\mu_o$ is the permeability of free space ($= 4\pi \times 10^{-7}$ H/m) and $\mu$ the relative permeability of the core material within the coil, such as iron. The reluctance is inversely proportional to the inductance, where $R = n^2/L$.

### Variable Reluctance Displacement Sensor

A **variable reluctance displacement indicator** uses a ferromagnetic material as its core, which usually is a semicircular ring. A small air gap separates the ends of the core from a plate. As the plate is moved over a surface, the air gap and, thus, the reluctance of the gap varies in time. This reluctance is in series with the core and plate reluctances. The total reluctance, $R_T$, becomes

$$R_T = R_o + kh,$$
(2.36)

in which $R_o$ depends upon core geometry and properties, $k = 2/(\mu_o\pi\ r_c)$, with $r_c$ the radius of the core material, and $h$ the air gap spacing (typically $\sim$1 mm). Using Equation 2.35, this leads to

$$L = \frac{L_o}{1 + kh/R_o},$$
(2.37)

in which the subscript $o$ denotes to zero-gap case, when $h = 0$. This equation shows that the inductance is inversely and nonlinearly related to the displacement. This limitation can be overcome by creating a displacement sensor that includes another inductance core that is positioned opposite the other and using a Wheatstone bridge containing the two inductors [3].

### Linear Variable Differential Transformer

The **linear variable differential transformer** (LVDT) measures displacements as small as $\sim$0.001 in. It operates on the principle of inductance, using a primary coil as the ac excitation source and two secondary coils. The secondary coils, configured in series, are placed on each side of the primary coil. Their output is an induced ac voltage. As a movable ferromagnetic core is displaced linearly from a center position in either direction, the rms of the induced voltage increases linearly to a point. Through signal conditioning of the rms voltages in the linear response region, the rms voltages in the negative direction are inverted such that the conditioned output voltage is linear with displacement, with negative voltage indicating displacement in one direction from the center position and positive voltage indicating the opposite.

### Reluctance Tachometer

The **reluctance tachometer** measures either linear or angular velocity. It operates, in principle, similar to the variable reluctance displacement sensor. The sensor is composed of a coiled wire with a stationary ferromagnetic core that is separated by a small air gap from either a linear translator or rotating wheel made of ferromagnetic, toothed material. The translator or wheel is attached to a moving part. As the teeth pass by end of the core, the air gap width changes. This alters the reluctance of the circuit, which manifests itself as a voltage induced in the coil. Using Equation 2.34, the induced emf can be expressed as

$$\mathsf{E} = -n\frac{d\phi_m}{d\theta}\frac{d\theta}{dt}. \tag{2.38}$$

The term $d\phi_m/d\theta$ varies periodically in time with a frequency corresponding to the tooth passage frequency. The term $d\theta/dt$ is the tooth passage frequency. Thus, both the amplitude and the frequency of the output voltage are proportional to either the linear or angular velocity. This sensor is used in a turbine flowmeter to measure volumetric flow rates.

## 2.4  Piezoelectric

**Piezoelectric material**, for example, quartz and polyvinyl difluoride (PVDF), has the property of developing an electric charge and, hence, an electric field when deformed. This is termed the **direct piezoelectric effect**. Conversely, when an electric field or voltage is applied to the material, it deforms. This is called the **converse piezoelectric effect**. The direct piezoelectric effect can be utilized in passive sensors, such as those for measuring displacement, stress, pressure, force, torque, acceleration, and sound. The converse piezoelectric effect can be used in active sensors, where application of a time-varying voltage produces a surface wave on the material. This effect can be utilized for sensing the presence of both inorganic and organic gases, as well as biological molecules [9].

A piezoelectric material, as its name implies, exhibits coupled mechanical and electrical behavior. When a mechanical stress is applied to a piezoelectric material, a strain occurs, thereby producing a displacement of ions within the stressed material. This induces or reorients the material's dipole moments. This yields two effects, an electric displacement (polarization) of the material and an electric field across the material. The polarization appears as a surface charge on the faces. The electric field results from the potential difference produced between the faces. Because of their atomic structure, piezoelectric materials have different electromechanical responses along each of the material's principal axes.

Typically, two opposing faces of the material have applied electrodes or a conducting material. The stress is applied perpendicular to the electrodes. For this situation, the surface charge, $q$, is

$$q = k_{pe} C_m F, \qquad (2.39)$$

in which $k_{pe}$ is a piezoelectric constant, $C_m$ the mechanical compliance, and $F$ the force applied over the face area, $A$.

Piezoelectric material with two plate electrodes is modeled electrically as a capacitor in parallel with a resistor. The capacitance, $C_o$, equals $\epsilon_o \epsilon A / h$, where $\epsilon$ is the relative permittivity of the material and $h$ is the distance between the electrodes. The resistance, $R_o$, equals $h/(SA)$, where $S$ is the electrode conductivity. The resistance is a consequence of the piezoelectric material having a finite conductivity (although it is a good dielectric) with the electrodes in series with the material having a much higher conductivity.

The capacitance acting in parallel with the resistance implies that the material has a time constant $(=R_o C_o)$. The time constant varies considerably with material, from 0.4 $\mu$s for lithium niobate to 2.2 h for quartz [9]. Hence, the output voltage on the electrode decays exponentially with time. This implies that some materials are better for static response measurements, such as quartz, and some are better for dynamic response measurements, such as lithium niobate.

Because $C_o = q/V_o$, where $V_o$ is the voltage on the electrode, Equation 2.39 implies that

$$V_o = (k_{pe} C_m / C_o) F. \qquad (2.40)$$

The quantity $k_{pe} C_m / C_o$ is called the sensor parameter, which varies approximately from 79.6 mV/N for lithium niobate to 582 mV/N for quartz [9]. This yields the piezoelectric coefficients of 2.3 pC/N for quartz and 6.0 pC/N for lithium niobate.

The output of a two-electrode piezosensor is charge. This and its relatively low capacitance requires that the output be conditioned prior to recording. If it is connected directly to a display device via a cable, then the cable and display introduce additional capacitances and resistances that reduce the time constant and sensor parameter considerably. The output can be conditioned using a charge amplifier to eliminate these deleterious effects. This, for example, increases the time constants to 50 ms for lithium niobate and 28 000 h for quartz [9].

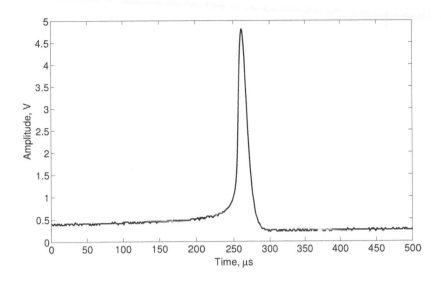

**FIGURE 2.7**

Charge detector output upon capture of a charged ($\sim$1 pC) ethanol micro-droplet [10]. Note the sensor's short time response, which was less than $\sim$100 $\mu$s.

The charge amplifier is an operational amplifier whose positive and negative inputs are the outputs of the two electrodes. A resistor is placed between one electrode and the positive input. A parallel resistor, $R_f$, and capacitor, $C_f$, are connected as a feedback loop between the amplifier's output and its negative input. In this manner, the current output of the sensor, which is $dq/dt$, is converted into an output voltage of the amplifier, $V_o$, because the current is integrated by the amplifier. This gives

$$V_o = -q/C_f. \tag{2.41}$$

Thus, the charge is linearly proportional to the voltage. An example oscilloscope tracing of the voltage output of a charge detector is shown in Figure 2.7. Here, the charge of a single ethanol micro-droplet was measured to be 10 pC. This **charge detector** can be used to measure charge as low as $\sim$1 fC under proper electrically shielded conditions [10].

**Piezoresistive Pressure Transducer**

A schematic of the cross-section of a miniature, integrated-silicon, piezoresistive pressure sensor is shown in Figure 2.8. A commercially available sensor is shown in Figure 2.9. Such sensors use a Wheatstone bridge containing the sensor and three other piezoresistors that are etched into a silicon diaphragm. Such pressure transducers often are used for unsteady

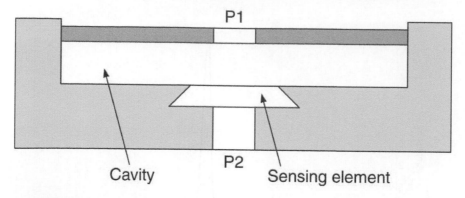

**FIGURE 2.8**
Schematic of an integrated-silicon piezoresistive pressure sensor.

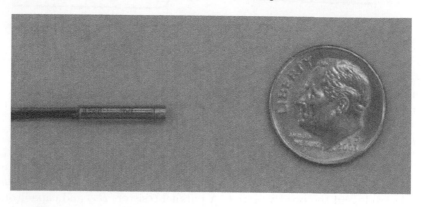

**FIGURE 2.9**
A commercially available piezoresistive pressure transducer. Note the size as compared to that of a U.S. dime.

pressure measurements because of their high-accuracy (typically less than 1 % of full scale) and high-frequency response (typically as high as ∼20 kHz). The sensor and its associated circuitry compose a **piezoresistive pressure transducer**.

### Piezoresistive Accelerometer

A **piezoresistive accelerometer** now is available [11] that contains a very small mass attached to a silicon cantilever beam instrumented with a piezoresistor. As the device is accelerated, the beam deflects, the piezoresistor is deformed, and its resistance changes. The piezoresistor is incorporated into a Wheatstone bridge circuit, which provides a voltage output that is linearly proportional to acceleration. The entire micro-accelerometer and associated circuitry is several millimeters in dimension.

## 2.5 Fluid Mechanic

The fluid mechanics of flow through a channel either with a variable cross-sectional area or with a constant cross-sectional area and an object placed within the flow can be exploited to determine either liquid or gas flow rates. Either type obstructs the flow in some manner. The choice of which flowmeter to use is based primarily upon the type of gas or liquid and upon the required accuracy.

### Pitot-Static Tube

Volumetric flow rate also can be determined by integrating measured velocity fields of the flow cross-section. Many different velocity sensors can be used for this purpose. In subsonic gas flows, the **Pitot-static tube** is used to determine flow velocity. This strictly is not a sensor but rather two concentric tubes that connect between the flow and at a pressure sensor. The center tube is open at its end and aligned axially into the flow. This is the total pressure port. The second tube is sealed on its end around the center tube. A short distance (typically, 3 to 8 tube diameters) from its sealed end, there are 4 holes at 90° intervals around the circumference whose axes are normal to the flow. These holes compose the static pressure port. Through Bernoulli's equation, the flow velocity, $U$, is determined as

$$U = \sqrt{\frac{2(p_t - p_s)}{\rho}}, \tag{2.42}$$

in which $p_t$ is the total pressure, $p_s$ the static pressure, and $\rho$ the density of the fluid. Uncertainties are introduced if the tube alignment is not directly into the flow (non-zero yaw angle) and at higher velocities (greater than $\sim35$ m/s in air).

There are a number of variations on this type of probe. The Pitot probe measures only total pressure. The Kiel probe has a shrouded inlet that makes it insensitive to yaw angle. The Pitot tube also can be used in supersonic gas flows, where the pressure rise across the shock wave that occurs at its inlet is accounted for using the the Rayleigh Pitot-tube formula.

### Variable-Area Flowmeters

The most commonly used flowmeters have variable cross-sectional areas. This common feature produces a pressure difference over the variable area. This pressure difference is related directly to the flow rate by using the conservation equations that govern fluid flow. These meters include, in order of increasing change in cross-sectional area, the **venturi**, the **flow nozzle**, and the **orifice plate**.

The conservation of momentum for an incompressible fluid with no change in elevation and a constant cross-sectional-area velocity can be expressed as

$$\Delta P = p_2 - p_1 = \frac{\rho U^2}{2}, \tag{2.43}$$

in which $p$ denotes the static pressure and the subscripts 1 and 2 the upstream and downstream positions, respectively. The conservation of mass for this case further implies that $Q = UA$. These two conservation equations can be combined to give

$$Q_i = \left[ \frac{A_2}{\sqrt{1 - (A_2/A_1)^2}} \right] \sqrt{\frac{2(p_1 - p_2)}{\rho}}, \tag{2.44}$$

in which $Q_i$ is an ideal flow rate, having no energy losses occurring in the flow as a result of the area change. The term in brackets in Equation 2.44 can be combined with discharge coefficient, $C$, to yield an expression for the actual flow rate, $Q_a$, where

$$Q_a = K A_2 \sqrt{\frac{2(p_1 - p_2)}{\rho}}, \tag{2.45}$$

in which $K = C/\sqrt{1 - (A_2/A_1)^2}$ is the flow coefficient and $C = Q_a/Q_i$. In practice, both $C$ and $K$ are different for each type of flowmeter. Thus, the flow rate can be determined from the specific values of these coefficients and the measured static pressure difference. The ranges of typical coefficient values, which are functions of the flow Reynolds number, are $0.95 \le C \le 0.99$ for the venturi, $0.92 \le C \le 0.99$ for the flow nozzle, and $0.60 \le K \le 0.75$ for a square-edge orifice.

### Laminar Flow Element

The **laminar flow element** (LFE) is designed to divide the flow cross-sectional area into a number of parallel channels, as many as $\sim 100$. The number of channels is dictated by maintaining laminar flow throughout each channel. This is ensured by keeping the channel Reynolds number, $Re_c$, less than $\sim 2300$, where $Re_c = \rho U_c d_c / \mu$, in which $\rho$ and $\mu$ are the density and absolute viscosity of the fluid, $U_c$ the fluid velocity in the channel, and $d_c$ the channel diameter. For a given fluid, this constrains the value of the product $U_c d_c$ because the flow rate through the channel, $Q_c$, given the total flow rate, $Q_a$, will equal $Q_c = Q_a/n$, where $n$ is the number of parallel channels.

Somewhat more complex flowmeters operate by placing an object in the flow. These include the rotameter, the turbine flowmeter, the magnetic induction flowmeter, and the vortex shedding flowmeter.

### Rotameter

The **rotameter** operates by having either a spherical or an elliptical weighted object (a float) contained in the flow. In its equilibrium position, the float's weight is balanced by its upward drag and buoyancy forces. The flow cross-sectional area is designed to increase linearly with upward distance and to achieve a float position that varies linearly with the flow rate, $Q$. The area variation with distance ensures that the annular velocity around the float, $U_a$, remains constant independent of float position. The flow rate is expressed as

$$Q = U_a A_a, \qquad (2.46)$$

in which $A_a$ is the annular area between the float and the rotameter's internal walls. Rotameters have visual scales and instrument uncertainties typically between 1 % and 10 %.

### Vortex-Shedding Flowmeter

There are several flowmeters that have objects placed in the flow that either rotate or vibrate with the flow. A **vortex-shedding flowmeter** has an object that extends across most of the flow channel and is connected by a beam to a force transducer located immediately outside of the channel. The cross-section of the object typically is square, semicircular, or triangular. The sectional dimensions of the object are chosen such that vortices are shed in a constant periodic manner as the fluid travels around the object. This shedding causes pressure to vary on the back of the object, which induces a period vibration in the object. This is sensed as a time-varying force by the transducer. Analytically, the volumetric flow rate is related to the shedding frequency, $f_s$, through the nondimensional Strouhal number, $St = 2\pi f_s D/U$, by the expression

$$Q = \frac{f_s D A}{2\pi St}, \qquad (2.47)$$

in which $D$ is the diameter of the object and $A$ the channel flow cross-sectional area. In this expression, it is assumed that the fluid velocity local to the object is the same as that over the flow area.

### Coriolis Flowmeter

Mass flow rate of either single-phase or multiphase fluids can be measured by a **Coriolis flowmeter**. The actual flow is routed through a U-shaped tube within the flowmeter. The tube is vibrated at a frequency, $\omega$, normal to the flow plane. As the fluid traverses one leg of the vibrating tube, it experiences a Coriolis force, $F_C$, where

$$F_C = 2\rho A \omega U L, \qquad (2.48)$$

in which $\rho$ is the fluid density, $A$ the flow cross-sectional area, $U$ the fluid velocity, and $L$ the length of one leg of the tube.

This force produces a torque that twists the tube and displaces it angularly several degrees out of the flow plane on both of its leg sides (one leg moves upward and the other leg downward). Thus, the total torque equals $2F_C h$, where $h$ is the lateral distance between the two legs of the tube. Noting that the mass flow rate of the fluid, $\dot{m} = \rho U A$, the mass flow rate can be related to the frequency by

$$\dot{m} = \frac{T}{2Lh\omega}. \qquad (2.49)$$

This can be expressed in terms of the twist displacement angle, $\theta_{tw}$, by noting that $\theta_{tw} = T/K$, where $K$ is the elastic stiffness of the tube. Thus, Equation 2.49 becomes

$$\dot{m} = \frac{K\theta}{2Lh\omega}. \qquad (2.50)$$

The twist displacement angle usually is small, such that $\sin\theta \simeq \theta \simeq \Delta y/(h/2)$, where $\Delta y$ is the vertical displacement normal to the plane of the tube with no vibration. This displacement can be measured using a variety of displacement sensors.

### Turbine Flowmeter

A **turbine flowmeter** has a rotor placed inside the flow. Its rotational frequency typically is sensed by an inductive pickup sensor that operates as a reluctance tachometer, as described before. For this flowmeter, the volumetric flow rate equals the product of a device-specific constant and the rotational frequency.

## 2.6 Optic

Measurement systems based upon optical sensing have been developed to measure velocity, temperature, and density. The actual sensor is just one part of the system that determines the measurand. Most optical systems are nonintrusive and active. They comprise a source of electromagnetic radiation, a medium that transmits and possibly alters the source radiation, and a detector that receives the radiation beyond the medium. In some, the medium changes the source radiation before it is detected, and in others, the source radiation is altered. A laser-Doppler velocimeter, for example, determines the velocity of a flowing medium by examining the frequency shift in laser light that is scattered by small particles moving in the medium. Here, the moving particles in the medium alter the characteristics of the source radiation. Alternatively, a pyrometer determines the temperature of a surface (the source of radiation) by measuring the intensity of the radiation received at a distance. Some of those used more commonly in thermomechanical systems are described in the following.

## Laser Doppler Velocimeter

The **laser Doppler velocimeter** is an optically based measurement system designed to measure noninvasively the velocity and velocity fluctuations of a transparent fluid over the velocity range from $\sim$1 cm/s to $\sim$500 m/s with $\sim$1 % accuracy. This system operates on the principle of the Doppler effect. The moving fluid is seeded with microparticles ($\sim$1 $\mu$m diameter), which ideally follow the flow.

*Single-Beam Method*: The **single-beam method** uses a monochromatic beam of laser light. For a fixed source of frequency $f_o$, the frequency observed by the moving particle, $f_p$, will be the Doppler shifted frequency, where

$$f_p = f_o \left[ \frac{1 - \frac{\vec{U} \cdot \hat{e}_o}{c}}{\sqrt{1 - (\frac{\vec{U} \cdot \hat{e}_o}{c})^2}} \right], \qquad (2.51)$$

in which $\vec{U}$ is the particle velocity, $\hat{e}_o$ the unit vector of the incident beam, and $c$ the speed of light. When $\vec{U} \cdot \hat{e}_o$ is much less than $c$, Equation 2.51 becomes

$$f_p = f_o \left[ 1 - \frac{\vec{U} \cdot \hat{e}_o}{c} \right]. \qquad (2.52)$$

Likewise, the frequency observed at the receiver (a photodetector), $f_r$, located in the same reference frame as the laser, will be the Doppler-shifted frequency of $f_p$, where

$$f_r = f_p \left[ 1 + \frac{\vec{U} \cdot \hat{e}_1}{c} \right], \qquad (2.53)$$

in which $\hat{e}_1$ is the unit vector of the scattered beam. Substitution of Equation 2.53 into Equation 2.52 yields the received frequency

$$f_r = f_o \left[ 1 + \frac{\vec{U} \cdot (\hat{e}_1 - \hat{e}_o)}{c} \right], \qquad (2.54)$$

in which the higher-order term is neglected. Thus, the total Doppler frequency shift will be

$$f_D = f_r - f_o = f_o \left[ \frac{\vec{U} \cdot (\hat{e}_1 - \hat{e}_o)}{c} \right] = \frac{|\vec{U} \cdot (\hat{e}_1 - \hat{e}_o)|}{\lambda}, \qquad (2.55)$$

in which $\lambda$ is the beam wavelength. Thus, with this method, only the velocity component in the $(\hat{e}_1 - \hat{e}_o)$ direction can be measured and not the total velocity. Further, because of the high frequency of the laser beam, the frequency of the scattered light that must be measured is very high.

**Example Problem 2.3**

*Statement:* The single-beam method is used to measure the stream-wise velocity component of air in a subsonic wind tunnel. The incident beam is argon-ion with a wavelength of 514.5 nm. Determine the frequency of the light that is scattered from a microparticle moving with the gas that is received by the photodetector.

*Solution:* First, assume that the velocity component in the wind tunnel is on the order of 10 m/s. The circular frequency of the incident argon ion beam, $f_{Ar-ion}$, equals $c/\lambda_{Ar-ion} = 3 \times 10^8/5.145 \times 10^{-7} = 5.83 \times 10^{14}$ rad/s. Noting that the frequency of the scattered light is the same as that received by the photodetector and using Equation 2.54, the frequency of the received light is

$$f_r \simeq 5.83 \times 10^{14} \left[ 1 + \frac{10}{3 \times 10^8} \right] \simeq 5.83 \times 10^{14}.$$

This frequency corresponds to $5.83 \times 10^{14}/2\pi = 92.8$ THz.

*Dual-Beam Method*: The **dual-beam method** overcomes the limitations of the single-beam method by using two beams of equal frequency, intensity, and diameter, and crossing the beams inside the flow. An interference pattern of fringes is formed as an ellipsoidal volume from the two incident circular beams. This volume has sub-millimeter dimensions [O(1 mm in length and 0.1 mm in diameter)].

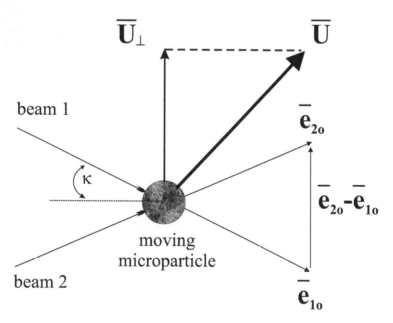

**FIGURE 2.10**
Dual-beam method geometry.

The incident light from each beam, after scattering from the microparticle as it moves through the ellipsoidal volume, is received by the same photodetector. Thus, applying Equation 2.55 to both beams gives a difference in the scattered frequencies as

$$f_{diff} = f_{r_2} - f_{r_1} = \frac{\left| \vec{U} \cdot (\hat{e}_{2_o} - \hat{e}_{1_o}) \right|}{\lambda}, \qquad (2.56)$$

in which $\hat{e}_{2_o} - \hat{e}_{1_o}$ is the difference in the unit vectors of the two incident beams, which is perpendicular to the ellipsoidal fringe spacing. This difference vector also is parallel to the component of the velocity, $U_\perp$, which is perpendicular to the bisector of the incident beams separated by an angle $2\kappa$, as shown in Figure 2.10.

Typically, $\kappa$ is small such that

$$\hat{e}_{2_o} - \hat{e}_{1_o} = 2\sin\kappa. \qquad (2.57)$$

Substitution of Equation 2.57 into Equation 2.56 yields

$$f_{diff} = \frac{2U_\perp \sin\kappa}{\lambda}. \qquad (2.58)$$

Thus, the difference in the scattered frequencies is linearly proportional to the velocity component perpendicular to the bisector of the two incident beams.

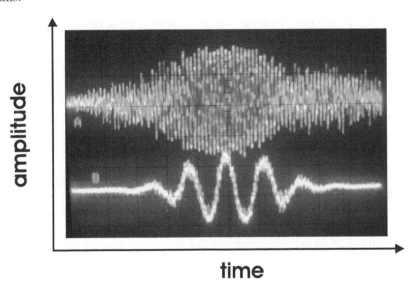

**FIGURE 2.11**
Doppler burst of a slowly moving water micro-droplet in air (top unfiltered; bottom filtered).

The interference pattern that is formed by the two intersecting beams has an intensity, $I_{int}$, that varies in the perpendicular bisector direction $x$ as $\cos(4\pi x \sin \kappa / \lambda)$. Thus, for one period of the intensity variation over the interference fringe spacing $d_f$, $4\pi d_f \sin \kappa / \lambda = 2\pi$, which reduces to

$$d_f = \lambda / 2 \sin \kappa. \tag{2.59}$$

The cyclic variation in intensity that is caused by the microparticle moving through the measurement volume is sensed by the photodetector and appears as a Doppler burst, as shown in Figure 2.11. Here, the time interval between cycles within the burst, $\Delta t$, using Equations 2.58 and 2.62, can be expressed as

$$\Delta t = \frac{d_f}{U_\perp} = \frac{\lambda}{U_\perp 2 \sin \kappa} = \frac{1}{f_{diff}}. \tag{2.60}$$

Thus,

$$U_\perp = \frac{\lambda f_{diff}}{2 \sin \kappa}, \tag{2.61}$$

in which $f_{diff}$ represents the observed Doppler frequency.

---

**Example Problem 2.4**

  *Statement:* The dual-beam method is used to measure the stream-wise velocity component of air in a subsonic wind tunnel. The incident beams are argon-ion with a wavelength of 514.5 nm. The beams are separated by a 20° angle. The time interval between cycles within a Doppler burst observed on an oscilloscope is 0.1 $\mu$s. Determine the fringe spacing in the measurement volume and the velocity component of the microparticle responsible for the burst.

  *Solution:* Using Equation 2.62, noting that $\kappa = 10°$, the fringe spacing is

$$d_f = 5.145 \times 10^{-7} / [(2)(0.174)] = 1.48 \ \mu\text{m}.$$

The velocity component is

$$U_\perp = d_f / \Delta t = 1.48 \times 10^{-6} / 1 \times 10^{-7} = 14.8 \ \text{m/s}.$$

---

Because either a positive or negative velocity component of the same magnitude will yield identical Doppler frequencies, frequency shifting by introducing a frequency difference of $\Delta f$ in one of the incident beams is employed. This leads to

$$f_{diff} = \Delta f + \frac{2\pi \vec{U} \cdot (\hat{e}_{2_o} - \hat{e}_{1_o})}{\lambda}, \tag{2.62}$$

which produces a signal for which $f_{diff}$ is less than $\Delta f$ for negative velocity components and greater than $\Delta f$ for positive velocity components.

Further modifications can be made by adding other beams of different frequencies in different Cartesian coordinate directions to yield all three components of the velocity.

### Phase Doppler Anemometer

For the **phase Doppler anemometer**, two additional photodetectors are added next to the existing photodetector of a dual-beam laser Doppler velocimeter. The three photodetectors are spaced apart equally by a small distance (several mm). When a microparticle passes through the measurement volume, this results in a Doppler burst that lags in time between one photodetector to the next. This phase lag can be related to the diameter of the microparticle using light-scattering theory. Thus, a three-component system can measure particle diameter in addition to its velocity components in all three orthogonal directions.

### Particle Image Velocimeter

One direct method to determine the velocity of a flow is to illuminate resident (either inherent or added) microparticles in the flow and track their position versus time. This method requires sufficient light scattering from the microparticles to illuminate them distinctly in the flow field. If successive images are obtained over short time intervals, realizations of the velocity field can be made. This method constitutes a **particle image velocimeter** (PIV).

Current PIV systems use a sheet of laser-light, generated using a plano-concave lens, to illuminate microparticles in a plane of the flow, as depicted in the top of Figure 2.12. Two successive images are obtained using a digital camera and a double-pulsed laser light having a precise time interval between the images. If the second light pulse occurs a short time after the first (usually $\mu$s to ms), then nearly instantaneous velocities in the plane of illumination can be obtained. Using subsequent sets of image pairs, the planar velocity field's velocity variations in time can be acquired. Various software packages are used to determine the planar velocity. One method correlates the light intensity in the center of a specific microparticle with intensities in the vicinity of where the microparticle should be in the subsequent frame (see Chapter 8 in [8] for a description of correlation functions). This is illustrated in the bottom of Figure 2.12.

### Particle Laser Light Sheet

Another alternative method for particle velocity determination is to illuminate the microparticle using a continuous laser light [12]. This is shown schematically in Figure 2.13. This is known as a **particle laser light sheet**. This light is chopped mechanically using a slotted rotating wheel, then expanded into a pulsed laser light sheet using a plano-concave lens. In this

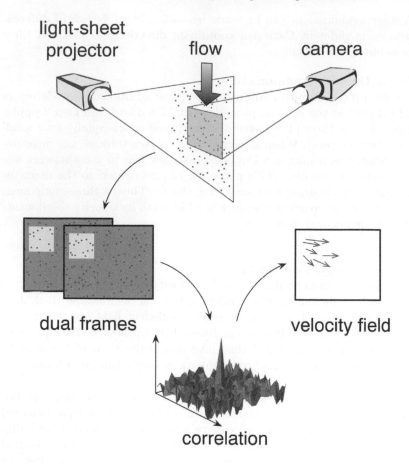

**FIGURE 2.12**
The PIV method.

manner, the wheel's radial velocity and slot width can be adjusted to yield a series of scattered light pulses. Here, the scaled length of the scattered light pulse divided by the pulse duration is the velocity in the light sheet plane. Although this method is easy to implement, it is less accurate than a conventional PIV system and requires a very dilute microparticle concentration to avoid overlapping scattered light pulses in an image.

**Transmitted and Reflected Light**

Light, either transmitted though or reflected from a multi-phase liquid-gas flow, can be used to determine important flow variables, such as bubble velocities and void fractions. This can be accomplished using records of successive ($\sim$10 $\mu$s image duration at $\sim$ 8000 images/s) images of the flow using a high-speed digital camera [13]. This approach is advantageous in that

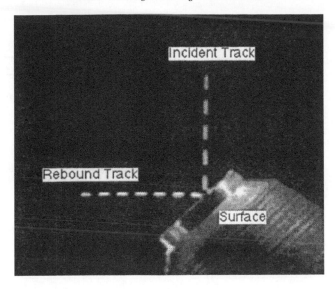

**FIGURE 2.13**
The strobed laser-light-sheet method. Vertical dashes are from a microparticle approaching the 45°-inclined surface; horizontal dashes from the microparticle rebound from the surface. More, shorter dashes correspond to a lower velocity.

it is noninvasive, compatible with most fluids, and able to distinguish clearly between gas and liquid phases. Further, the effective frequency response of this approach can be greater than 10 MHz.

The void fraction, $\alpha$, for a liquid-gas mixture is defined as

$$\alpha \equiv V_g/(V_g + V_l), \tag{2.63}$$

where $V_g$ is the gas volume and $V_l$ is the liquid volume. Time and space-averaged pixel intensities of images of the $x, y$ directional plane of the flow and their projection into the $z$ direction (depth) of the flow can be used to determine the local gas and liquid volumes, and, thus, the void fraction. The contrast between the gas and liquid phases can be increased by back-lighting the flow with a diffuse light source. Dark regions correspond to the gas phase and light regions to the liquid phase. An example image that illustrates the results of phase contrasting is presented at the top of Figure 2.14.

Image analysis to determine the void fraction can be performed by calculating the time-averaged pixel intensity at a particular $x, y$ location within a square interrogation region of size 3 pixels by 3 pixels ($\Delta x = 13.7$ $\mu$m by $\Delta y = 13.7$ $\mu$m). The dimensions of the interrogation region usually are chosen to be smaller than the smallest gas void observed, thus ensuring that either gas or liquid completely occupies the interrogation region.

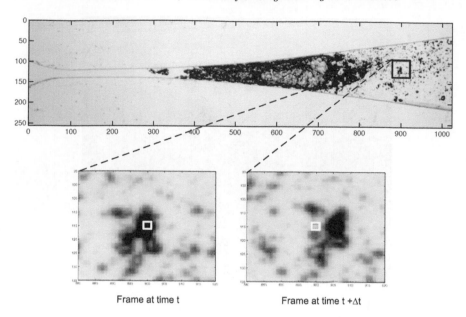

Frame at time t                    Frame at time t +Δt

**FIGURE 2.14**

Top: 1024 pixel ($x$ direction) by 256 pixel ($y$ direction) digital image of cavitating water in a nozzle. Flow is from left to right. Bubbles appear as dark structures; the liquid phase as bright regions. Bottom: Two sequential magnified images of the black-bordered region of the flow in which the interrogation region, centered at the pixel coordinates (900,115), is indicated by the white-bordered region. The frame rate was 8000 frames/s with a shutter exposure duration of 6.7 $\mu$s.

The spatial average intensity of all the pixels within the interrogation region at a particular time, $I(t, x, y)$, is determined. A threshold filter can be applied to compensate the rare instances when all 9 pixels do not have the same intensity. This approach results in the phase scalar, $X(x, y, t)$, determined by the conditions

$$X(x,y,t) = \begin{cases} 1 & \text{if I(x,y,t)} > \text{P} \\ 0 & \text{if I(x,y,t)} < \text{P} \end{cases}, \tag{2.64}$$

where $P$ is a threshold pixel intensity used to distinguish between the two phases. Thus, for a given time and $x, y$ location, the phase is considered to be either gas ($X(x, y, t) = 1$) or liquid ($X(x, y, t) = 0$) within the interrogation region.

Two sequential, magnified images of a region of the flow are shown in the bottom of Figure 2.14. For the left image, a dark gas void covers the interrogation region contained within the white-bordered box. $I(x, y, t)$ is greater than $P$ and, thus, $X(x, y, t) = 1$. For the right image, which is the next image in the sequence, the gas void has convected downstream and liquid occupies the interrogation region outlined by the white-bordered box. $I(x, y, t)$ is less than $P$ and, thus, $X(x, y, t) = 0$.

The phase scalar $X(x, y, t)$ is averaged over time to yield the time-mean phase scalar, $\bar{X}(x, y)$, given by the expression

$$\bar{X}(x, y) = \frac{1}{T} \int_0^T X(x, y, t) dt, \tag{2.65}$$

where $T$ is the total time record length of a sequence of images. $T$ is chosen to be long enough to achieve a sufficiently converged value of $\bar{X}(x, y)$, typically ~500 frames, corresponding to $T \simeq 34$ ms.

The time-mean phase scalar is subsequently integrated in the $y$ (cross-stream) direction over the local height of the nozzle, $h(x)$, to yield the mean phase scalar at that $x$ location, $\bar{X}(x)$, defined as

$$\bar{X}(x) = \frac{1}{h(x)} \int_0^{h(x)} \bar{X}(x, y) dy. \tag{2.66}$$

This phase scalar represents the spatial-temporal average of the ratio of area of the gas to the total area, where the total area equals $\Delta x \cdot h(x)$.

Finally, the void fraction at a given $x$ location is determined as

$$\alpha(x) = [\bar{X}(x)] \cdot [\bar{X}(z)]. \tag{2.67}$$

In this expression,

$$\bar{X}(z) = \frac{1}{d_n} \int_0^d X(z) dz, \tag{2.68}$$

where $\bar{X}(z)$ is the mean phase scalar for the $z$-direction (nozzle depth) and $d_n$ denotes the nozzle depth.

The overall uncertainty in the void fraction is less than 10 %. The uncertainty in the bubble velocity as estimated from high-speed digital photography is less than 5 %.

### Pyrometer

Pyrometry refers to measuring body surface temperatures greater than ~750 K. Above this temperature a body will emit electromagnetic energy in the visible range. All **pyrometers** are based upon the principle that a body emits an amount of electromagnetic energy that depends upon its wavelength and the surface temperature of the body. The spectral intensity distribution, $I_{\lambda, T}$, known as the Planck distribution, is expressed as [14]

$$I_{\lambda, T} = \frac{\epsilon A_o}{\lambda^5 (e^{A_1/\lambda T} - 1)}, \tag{2.69}$$

in which $T$ is the absolute temperature in K, $\lambda$ is the wavelength in $\mu$m, $\epsilon$

is the emissivity, and $A_o = 2\pi\hbar c_o^2$ and $A_1 = \hbar c_o/k_B$ are constants equal to 374.15 MW $\mu m^4/m^2$ and 14 388 $\mu m$ K, respectively. The units of $I_{\lambda,T}$ are W/(m$^2$ $\mu m$). Planck's constant, $\hbar$, equals $6.6256 \times 10^{-34}$ Js and Boltzmann's constant $k_B$, equals $1.3805 \times 10^{-23}$ J/K. When $\epsilon = 1$, the body is termed a blackbody or ideal radiator. When $\epsilon < 1$, the body is termed a non-ideal radiator.

When a body is in thermal equilibrium with its surroundings, it emits the same amount of energy that it absorbs. In this case, the amount of emitted energy as compared with the total energy, which also includes reflected and transmitted energies, is characterized by its emissivity. Surfaces such as paper, brick, smooth glass, and water have $\epsilon > 0.9$, whereas polished metals have $\epsilon < 0.1$.

The heat flux between a non-ideal body, A, and a perfectly absorbing body, B, $q$, is

$$q = \epsilon A F_{AB}\sigma(T_A^4 - T_B^4),\tag{2.70}$$

in which $F_{AB}$ is the shape factor that depends upon the geometry of the bodies A and B, and $\sigma$ is the Stefan-Boltzmann constant (=5.6697 $\times$ 10$^{-8}$ W/(m$^2$ K$^4$). The heat flux equals the integral of the spectral intensity distribution (given by Equation 2.69) over the appropriate wavelength range. The consequent expression relates heat flux, intensity, wavelength, and temperature.

There are several types of optical pyrometers. Each of them receive the emitted energy, some at one, some at two, and some over a range of wavelengths. The **optical pyrometer** measures the source intensity at one wavelength, typically in the red. A **two-color pyrometer** measures the intensity at two different wavelengths. From Equation 2.69, it can be seen that the effect of emissivity can be eliminated when using the ratio of the two measured intensities. The **infrared pyrometer** measures intensities over a range of infrared wavelengths, and the **total-radiation pyrometer** measures intensities over a very wide range of wavelengths. Either photon or thermal detectors are used as photodetectors to receive the radiated energy. Each type of pyrometer is constructed and operates differently.

### Nondispersive Infrared Detectors

The concentrations of certain gases, such as CO, $CO_2$, and $SO_2$, can be measured using the principle of the absorption of electromagnetic radiation by a molecule as a function of the wavelength of the incident radiation. This is expressed through the Beer-Lambert law [15] as

$$I = I_o e^{-\epsilon_m L[C]},\tag{2.71}$$

in which $I$ is the received intensity over the wavelengths of interest, $I_o$ the reference intensity, $\epsilon_m$ the molar absorptivity, $L$ the path length over which the absorption occurs, and [C] the molar concentration. In particular,

$CO$, $CO_2$, $SO_2$, and water vapor absorb radiation in the infrared band very effectively, as opposed to $O_2$ and $N_2$.

The **nondispersive infrared detector** operates by having two identical length and material cells, one reference cell filled with a nonabsorbing gas and the other through which a dehumidified sample gas is passed. A beam of infrared light is passed consecutively through each cell using a beam chopper. The concentration of a particular species is related by Equation 2.71 directly to the ratio of the two intensities measured using photodetectors.

**Optical Density Methods** Various optical methods have been developed for use in gases that exploit the relation between the refractive index of a gas with its density. These include the interferometer and the Schlieren. They are based upon and can determine density and its first spatial derivative, respectively. Each of these methods are explained in the following.

These methods are based upon several fundamental and related principles. The speed of light varies with the density of a medium as

$$\rho = \rho_{STP} \frac{c_o - c}{\beta c_o}, \tag{2.72}$$

in which $\rho_{STP}$ is the density of the medium at standard temperature and pressure, $c_o$ the speed of light *in vacuo* ($= 2.998 \times 10^8$ m/s), $c$ the speed of light in the medium, and $\beta$ a constant for the medium ($= 0.000\ 292$ for air at standard temperature and pressure). The speed of light in the medium is related to the speed of light *in vacuo* through the index of refraction of the medium, $n$ by

$$c = c_o/n. \tag{2.73}$$

For light traveling in the positive $x$ direction across a test section of width $W$ perpendicular to the flow in the $z$ direction, a light ray will be turned gradually through an angle $\Delta\alpha$ in the negative $y$ direction if the density increases in that direction. Thus,

$$\Delta\alpha = \frac{W}{c} \frac{dc}{dy}. \tag{2.74}$$

Equation 2.73 can be substituted into Equation 2.74 to yield

$$\Delta\alpha = -\frac{W}{n} \frac{\beta}{\rho_{STP}} \frac{d\rho}{dy}. \tag{2.75}$$

Thus, the turning angle is linearly proportional to the density gradient that occurs over the test section width.

**Interferometer**

The **interferometer** uses a monochromatic source of light that is optically split and passed as two beams through a test section containing a gas. The beams exiting the test section are focused upon a screen or CCD

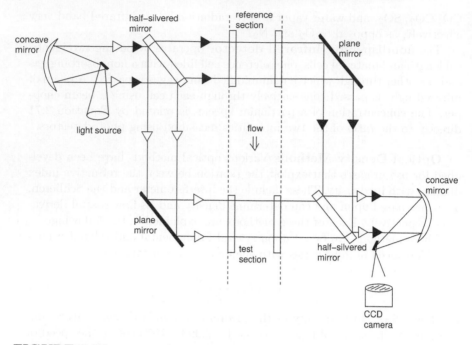

**FIGURE 2.15**
The Mach-Zehnder interferometer.

(charge coupled device) array using a lens as shown in Figure 2.15. If there are no density variations along each beam path, then the two beams will constructively interfere and form a pattern of parallel light fringes and dark bands. If one of the beams passes through a density region that is different from the other beam, the fringes will be displaced by a distance $N\lambda$ from their reference position, where $N$ is the number of fringes and $\lambda$ is the wavelength of the monochromatic light.

When the index of refraction of a gas increases from $n_o$ to $n_m$, where the subscripts $o$ and $m$ denote the stagnant and moving gases, respectively, the number of fringes over which a shift in the fringe pattern occurs will be

$$N = W(n_m - n_o)/\lambda, \tag{2.76}$$

in which $W$ is the width of the test section. Using Equations 2.72 and 2.73, Equation 2.76 becomes

$$N = \beta \frac{W}{\lambda} \frac{\Delta\rho}{\rho_{STP}}, \tag{2.77}$$

in which $\Delta\rho = \rho_m - \rho_o$. Thus, the change in density, $\Delta\rho$, for a given set of test section, gas and light source conditions, is proportional to the number of fringes over which a shift in the fringe pattern occurs. Using this information and the reference density, the density field of the moving gas can be determined. Further, if the field of another intensive thermodynamic quantity of the gas is known, such as pressure, then the fields of other thermodynamic quantities can be found, such as temperature, enthalpy, and entropy.

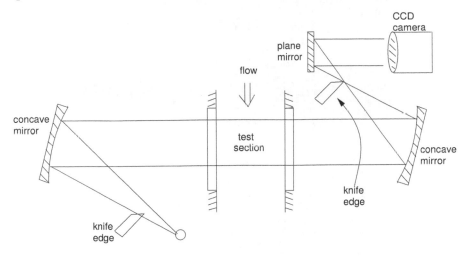

**FIGURE 2.16**
The Schlieren.

### Schlieren

The **Schlieren**, as depicted in Figure 2.16, uses a light line source whose focal point is in the plane of a knife edge. This makes the light-source image parallel to the knife edge. Approximately one-half of the light is blocked by the knife edge. This produces a sharper light beam, which is directed to a concave mirror. This mirror transmits the beam through the test section and to a second concave mirror. A second knife edge is placed parallel to the first and in the plane of the focal point of the second mirror. The transmitted beam is then reflected off of a plane mirror and projected unto a screen or a CCD array. A thin knife edge is positioned at the focal length of the second lens, $L_2$, and at a small distance $D$ in the negative $y$ direction below the optical axis. The knife edge is used to control the amount of illumination on the screen by blocking some of the transmitted light. This system is sensitive to density gradients and, thus, is used to visualize regions of large density gradients, such as shock waves. An example of an acquired image is presented in Figure 2.17.

For a beam of light refracted through the test section by an angle $\alpha$ such that $\alpha = D/f$, where $f$ is the focal length of the second concave mirror, using Equation 2.75 gives

$$D = -f\frac{W}{n}\frac{\beta}{\rho_{STP}}\frac{d\rho}{dy}. \tag{2.78}$$

**FIGURE 2.17**
Schlieren image of supersonic flow ($M_\infty = 2$) over an inclined wedge. The sharply contrasted light-to-dark line above the darkwedge corresponds to an oblique shock wave. The light line below the wedge is a weak expansion fan. Image courtesy of Prof. S. Gordeyev, University of Notre Dame.

This equation can be integrated to yield

$$\rho(y) = \rho_{ref} + \frac{n\rho_o}{fW\beta} \int_{y_o}^{y} D(y)dy, \qquad (2.79)$$

in which $\rho_{ref}$ is a reference density at some point in the flow. This equation gives density as a function of $y$ position. Because the relative change in illumination is proportional to the density gradient at a fixed knife edge position, the relative intensities of the illumination can be viewed as density variations.

## 2.7 Photoelastic

Methods using transmitted light can be used to determine stress and strain fields. The photoelastic method determines stress fields in certain transparent materials, such as plastics or materials coated with a photoelastic layer. The moiré method determines the strain field in a material and requires a grating material to be fixed to the material's surface. Both methods rely upon developing an interference pattern consisting of bright and dark regions.

The **photoelastic method** is based upon the principle that certain materials will polarize light when they are strained. More specifically, if light is passed through a material that is strained in one direction, the material becomes doubly refracting, establishing two polarizing planes with different refractive indices that are orthogonal to the axis of the applied stress.

### Plane Polarizer

A **plane polarizer** contains two polarizing lenses. The specimen is placed between the two lenses. Light from a source is plane polarized by the first polarizer. This light enters the material under stress, which produces two planes of polarized light, one along each principle-stress axis and one phase-lagged with respect to the other. This light then is plane polarized by a second polarizer, retaining two waves that interfere with one another because of their phase difference.

Quantitatively, the relative phase retardation, $\Delta\phi$, is proportional to the difference in the indices of refraction in the directions of the principle axes, $n_{11} - n_{22}$. This difference, further, is proportional to the difference in stresses, $\sigma_{11} - \sigma_{22}$, along the principle axes. Thus,

$$\Delta\phi = Ct(\sigma_{11} - \sigma_{22}), \tag{2.80}$$

in which $C$ is the stress optic coefficient and $t$ the material thickness along the polarizer axis. The extent of the fringe pattern that is produced is related to $\Delta\phi$ and, hence, the difference in the principle stresses. Calibration of the polarizer using a material with known properties in simple tension is required to determine the magnitudes of the stresses throughout the material. As a load is applied gradually to the calibration material, each developed fringe can then be associated with a value of calculated stress. Thus, the order of the fringes can be related directly to stress levels. Then, when stress is applied to a test material, the stress field can be determined from a series of images obtained at different orientations of the material with respect to the polarizer axis.

### Moiré

The **moiré method** uses the fringe pattern that is formed by two overlapping and similar gratings. The grating formed consists of equally spaced dark lines (actually dark bands of a given width, with up to more than 1000 lines/mm possible). One grating, the working grating, is deposited on the surface of the material. The other, the reference grating, is placed on the surface of the material in contact with the working grating.

As the material is strained, the lines of the working grating are displaced with respect to the reference grating. Also, the widths of the bands in the working grating increase slightly. The relative shift in lines produces regions of different intensities that vary from light to dark. The center of a darkest region can be considered to be the center of a fringe.

Assume that the reference grating has a pitch (the distance between lines) of $p_r$ and the working grating an initial pitch $p_r$ that becomes $p_r + \delta p$ under strain, with $0 < \delta p < p_r$. The distance from an arbitrary line on the reference grating, $d_r$, to its $n$-th line will equal $np_r$. Likewise, the distance to the working grating's $n$-th line will equal $n(p_r + \delta p)$. So the relative distance between the $n$-th working grating line and the $n$-th reference grating line is $n\delta p_r$.

Generalizing the above by letting $n_r$ correspond to $n$ of the reference grating, $n_w$ to that of the working grating, and $\alpha = \delta p/p_r$, the expressions for $d_r$ and $d_w$ become

$$d_r = n_r p_r \tag{2.81}$$

and

$$d_w = n_w(p_r + \alpha p_r). \tag{2.82}$$

Fringes will be formed when the lines from both gratings overlap. This is when the distances $d_w$ equals $d_r$ or an integer multiple of it, noting that the value of $n$ for the reference grating will be different than the $n$ for the working grating. Equating Equations 2.81 and 2.82 gives

$$\frac{n_w}{n_r} = \frac{p_r}{p_r + \alpha p_r} = \frac{1}{1 + \alpha}. \tag{2.83}$$

For example, when $\alpha = 0.5$, $n_w/n_r = 2/3$. This implies that the first fringe will occur when $n_w = 2$ and $n_r = 3$. That is, the second working grating line will overlap the third reference grating line to form the first fringe. Because fringes are equally spaced for the same strain, the second fringe will occur at $n_w = 4$ and $n_r = 6$, and so forth. In general, the first fringe will occur when the $n_w = N_l$ and $n_r = D_l$, where $N_l$ is the lowest integer value of the numerator and $D_l$ is the lowest integer value of the denominator in the fraction $1/(1+\alpha)$.

There are two important consequences of the fringe spacing. First, the fringe spacing can be directly related to the strain using the aforementioned procedure to determine $\alpha$. Second, the pitch of the reference grating is related to the fringe spacing through $\alpha$. In fact, the fringe spacing equals

$pD_l$. Finally, the stress field can be computed from the strain field using the material's stress-strain relation.

## 2.8   Thermoelectric

When wires consisting of two dissimilar metals are joined at each of their two ends and the ends are exposed to different temperatures, an emf will be generated in an open circuit formed by having a break in one of the wires. This emf varies with the type of metal and the temperature difference. This phenomena is known as the Seebeck effect. This emf results from two possible emfs, one occurring at each junction with its different temperature (the Peltier effect) and the other along each metal because of its temperature gradient (the Thomson effect). Usually, the emf generated by the Thomson effect is negligible with respect to the Peltier-effect emf.

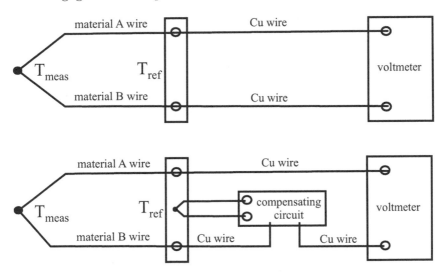

**FIGURE 2.18**
Simple thermocouple circuit (top) and one using an electronic reference junction (bottom).

This effect forms the basis of the **thermocouple**. An actual thermocouple (TC) is simply one junction (the *hot* junction) and its two metal wires. This is connected to a circuit that contains the other junction (the *cold* junction). The cold junction serves as a reference junction that is held at a constant temperature, usually 0 °C. One of the wires is interrupted between the two junctions by connecting copper wires to the ends, which are at a constant temperature, and then connecting the copper wires to a high-impedance voltmeter or amplifier. An alternative arrangement is to connect each of both metals to a copper wire at a constant-temperature junction and then connect the copper wires to a voltmeter, as shown in the top of Figure 2.18. Both arrangements produce the same emf.

Today, electronic reference junctions for each type of metal pairs are available. The reference junction temperature usually is measured by a thermistor. Most reference junctions also come with built-in amplifiers and linearizers, all embedded on one electronic chip. Because the direct output voltages of thermocouple circuits at the mV level, high-input impedance meters are required to measure the voltage (see Chapter 2 of [8]). This problem, however, is circumvented when using the electronic reference junction/amplifier/linearizer chip.

Each pair of metals used for thermocouples have different sensitivities. The chromel/constantan pair has the highest sensitivity ($\sim 40$ $\mu V/°F$). The platinum/platinum-10 % rhodium pair has the lowest sensitivity ($\sim 8$ $\mu V/°F$). All pairs exhibit non-linear sensitivities within the approximate range from 0 °F to 300 °F and are linear for higher temperatures. Standard calibration curves of mV versus T are available from the NIST [16].

Thermocouples have instrument uncertainties less than 1 %. Their response times, which typically range from several ms to several s, depend upon the environment in which they are used and their diameter (see Chapter 4 in [8] for a model of thermocouple response). Thermocouples also can be arranged in series to measure temperature differences, in parallel to measure average temperature, and in multiple junctions (thermopiles) to amplify the output.

## 2.9     Electrochemical

Electrochemical sensors are the most similar to those within the human body that are described in Chapter 3. Both types have an induced chemical change that leads to an electrical response. Electrochemical sensors can be subdivided into different categories based upon how they respond to chemical stimuli [9]. **Conductimetric** sensors respond to a chemical stimulus by a change in either resistance or impedance. Many have metal-oxide-semiconductor (MOS) layers deposited on the sensing surface. **Chemiresistors** are conductimetric sensors with organic or metallic sensing surfaces. **Potentiometric** sensors respond with a change in potential (voltage) and **amperometric** sensors with a change in current. Modern examples of electrochemical sensors include the Taguchi sensor (conductimetric) that measures gas concentration, the air/fuel sensor (conductimetric) in automobiles that measures oxygen partial pressure, and the chemically sensitive field-effect transistor (ChemFET) sensor (amperometric) that can detect atoms, molecules, and ions in liquids and gases through the charge transfer that occurs between the chemical and the sensing material. Electrochemical sensors can sense material in the solid, liquid, or gas phase. They have been used to measure gas concentration, pH, humidity, and various biological molecules.

### Conductimetric

Conductimetric sensors are composed of a substrate insulator upon which a semiconductor or metallic sensing film is deposited along with end electrodes. A potential difference between the electrodes establishes a current flow through the sensing material. As that material is exposed to a chemical stimulus, its carrier concentrations and mobilities change, which effectively changes its resistance (in the case of a dc potential) or inductance (in the case of an ac potential). This change can be related to the level of the stimulus. The change can be transduced a number of ways to yield a detectable voltage output, as described previously for resistive output sensors. The frequently used Taguchi sensor is based upon this principle [9]. This sensor uses a $SiO_2$ material and can measure the concentrations of combustible and reducing gases.

### Potentiometric

Potentiometric sensors are based upon generating an electrochemical emf from the exchange of electrons between a species in solution and the sensing element.

**Ion selective electrodes** are used to sense the activity or concentration of ions in solution. The ions of the species in solution electrochemically react with those species on the electrode. This establishes an electrode potential, $E$, determined by the Nernst equation as

$$E = E_o - 2.303 \frac{\mathcal{R}T}{n_e F} \log_{10}(a_A + \beta_{AB} a_B), \qquad (2.84)$$

in which $E_o$ is a constant that depends upon the electrode material in V, $\mathcal{R}$ is the universal gas constant equal to 8313.3 J/(kg-mole·K), T is the absolute temperature in K, $n_e$ is the number of electrons involved in the reaction, $F$ is Faraday's constant equal to 96 485 C, $a_A$ is the activity of species A, $a_B$ is the activity of species B, and $\beta_{AB}$ is the selectivity coefficient of an A species-sensitive electrode to species B.

The activity of a species A, $a_A$, is related to its activity coefficient, $\gamma_A$, and its concentration, $[C_A]$ by

$$a_A = \gamma_A [C_A], \qquad (2.85)$$

in which $\gamma_A < 1$. Because the activity depends, to some extent, on the concentrations of all other ions and species in solution, $\gamma_A$ is a non-linear function of $[C_A]$. When $[C_A]$ is less than $\sim 0.1$ mmole/L, $\gamma_A$ is approximately unity. There, $a_A \simeq [C_A]$. In either situation, Equation 2.85 can be substituted into Equation 2.84 to relate electrode potential to the concentration of a species in solution.

This method can be used to measure the concentration of hydrogen ions in solution. This, in turn, determines the pH of the solution through the expression

$$pH = -\log_{10} a_{H^+}. \tag{2.86}$$

An electrochemical potential also can be established across a membrane that is permeable to particular molecule, such as zirconia oxide to oxygen and palladium to hydrogen. This forms the basis of certain gas sensors.

### Amperometric

The ChemFET sensor is one example of an amperometric sensor. This sensor behaves similarly to an electrical circuit MOSFET transistor [17]. As a chemical stimulus is applied to this sensor, a charge transfer occurs between the chemical and the sensor's gate electrode. This changes the drain current, which manifests itself as a change in the threshold voltage. Thus, either the drain current or the drain current-threshold voltage relation can be used to determine the chemical concentration.

## 2.10 Problems

1. A metallic wire embedded in a strain gage is 4.2 cm long with a diameter of 0.07 mm. The gage is mounted on the upper surface of a cantilever beam to sense strain. Before strain is applied, the initial resistance of the wire is 64 $\Omega$. Strain is applied to the beam, stretching the wire 0.1 mm, and changing its electrical resistivity by $2 \times 10^{-8}$ $\Omega$m. If Poisson's ratio for the wire is 0.342, find the change in resistance in the wire due to the strain to the nearest hundredth ohm.

2. A NTC thermistor with has Steinhart-Hart coefficient values of $A = 1.5 \times 10^{-3}$ K$^{-1}$, $B = 2.5 \times 10^{-4}$ K$^{-1}$, and $C = 1 \times 10^{-7}$ K$^{-1}$. Its measured resistance is 2.5 k$\Omega$ when immersed in a liquid. Determine the temperature (in K) of the liquid.

3. A metallic strain gage has a resistance of 350 $\Omega$ at zero strain. It is mounted on a 1-m-long column. The column is strained axially by 1 cm. Determine a typical resistance (in $\Omega$) of such a gage under its strained condition.

4. A resistive accelerometer is fabricated with an internal mass of 1 gm and 2-mm-long strain gages, each having a spring constant of 300 N/m. When the gages are strained by 2 % in a direction parallel to the strain gages, determine (a) the acceleration (in m/s$^2$) in the direction parallel to the strain gages and (b) the acceleration (in m/s$^2$) in the direction perpendicular to the strain gages.

5. A variable-capacitance relative humidity sensor has a capacitance of 10 $\mu$F at 10 % relative humidity and 35 $\mu$F at 50 % relative humidity. Determine (a) its capacitance at 78 % relative humidity, (b) its capacitance at 0 % relative humidity, and (c) its sensitivity.

6. (a) Determine the total pressure-minus-static pressure difference (in Pa) that is measured by a pressure transducer connected to a Pitot-static that is positioned in a wind tunnel where the velocity of the air is 30 ft/s. (b) Express this pressure difference in units of atm and in units of in. H$_2$O. Finally, (c) identify a pressure transducer model and manufacturer that would be suitable.

7. The Strouhal number, $St$, depends only on the Reynolds number, $Re$. For a cylinder in cross-flow, $St$ is constant and equals 0.21 for 6000 $\leq Re \leq$ 60 000. For a vortex shedding flowmeter using a 1-cm-diameter cylindrical element placed in water under standard conditions in this $Re$ range, determine the range of shedding frequencies (in Hz).

8. Determine the emf generated by a 20-turn 0.6 mH inductor when the ac current varies $3\sin(120\pi t)$ mA/s.

9. Lithium niobate is used as the active material in a piezoresistive pressure transducer. Its sensing area is 1 cm$^2$. If the transducer is exposed to a pressure of 2 atm, determine its (a) electrode voltage (in V) and (b) surface charge (in C).

10. Referring to the strobed laser-light-sheet image that is shown in Figure 2.13, (a) determine the ratio of the microparticle's rebound velocity to its incident velocity. Noting that the normal coefficient of restitution, $e_n$, is defined as the ratio of the normal component of the rebound velocity to the normal component of the incident velocity, (b) determine $e_n$.

11. The same nondispersive infrared detector is used to measure the concentrations of $CO_2$ for fuel-rich and fuel-lean conditions. Determine the ratio of the fuel-rich to fuel-lean $CO_2$ concentrations when the measured intensities are 51 W/m$^2$ and 8 W/m$^2$, respectively, and the reference intensity, $I_o$, is 2 W/m$^2$.

12. The supersonic flow of air over a 15° wedge at $M = 1.7$ produces an oblique shock wave. Through the shock wave, the density increases by 174 %. A Mach-Zender interferometer is used with a source wavelength of 530 nm. The test section width is 4 in. Determine (a) the number of fringe shifts that occurs through the shock and (b) the change in the index of refraction through the shock.

13. The moiré method is used to determine the strain of a material. The reference grating has a pitch of 500 lines/mm. Assume $\alpha = 0.2$. Upon strain, the third fringe is observed to occur where the 15th working grating line overlaps the 18th reference grating line. Determine the strain for these conditions.

14. Determine the approximate emf (in mV) that is generated by a chromel/constantan thermocouple exposed to a 350 K gas and referenced to 273.15 K.

15. A hydrogen ion selective electrode measures a pH of 5.0. Determine the concentration (in mmole/L) of the hydrogen ions in solution.

# Bibliography

[1] Alciatore, D.G. and Histand, M.B. 2003. *Introduction to Mechatronics and Measurement Systems*. 2nd ed. New York: McGraw-Hill.

[2] Beckwith, T.G., Marangoni, R.D. and Leinhard V, J.H. 2006. *Mechanical Measurements*. 6th ed. New York: Addison-Wesley.

[3] Bentley, J.P. 2005. *Principles of Measurement Systems*. 4th ed. New York: Pearson Prentice Hall.

[4] Wheeler, A.J. and A.R. Ganji. 2009. *Introduction to Engineering Experimentation*. 3rd ed. New York: Prentice Hall.

[5] Doebelin, E. 2003. *Measurement Systems: Application and Design*. 5th ed. New York: McGraw-Hill.

[6] Figliola, R. and Beasley, D. 2010. *Theory and Design for Mechanical Measurements*. 5th ed. New York: John Wiley and Sons.

[7] Lykoudis, P.S. and Dunn, P.F. 1973. Magneto-Fluid-Mechanic Heat Transfer from Hot-Film Probes. *Int. J. Heat and Mass Trans.* 16, 1439–1452.

[8] Dunn, P.F. 2010. *Measurement and Data Analysis for Engineering and Science*. 2nd ed. Boca Raton: CRC Press; Taylor and Francis Group.

[9] Vetelino, J. and Reghu, A. 2011. *Introduction to Sensors*. New York. CRC Press.

[10] Kim, O.V. and Dunn, P.F. 2010. Real-Time Direct Charge Measurements of Microdroplets and Comparison with Indirect Methods. *Aerosol Sci. & Tech.*, 44, 292–301.

[11] Hsu, T.-R. 2002. *MEMS & Microsystems: Design and Manufacture*. New York: McGraw-Hill.

[12] Dunn, P.F., Brach, R.M., and Caylor, M.J. 1995. Experiments on the Low Velocity Impact of Microspheres with Planar Surfaces. *Aerosol Sci. and Tech.* 23, 80–95.

[13] Dunn, P.F., Thomas, F.O., Davis, M.P. and Dorofeeva, I.E. 2010. Experimental Characterization of Aviation-Fuel Cavitation. *Phys. Fluids*, 22, 117102-1–117102-17.

[14] Incropera, F.I. and De Witt, D.P. 1985. *Fundamentals of Heat and Mass Transfer.* 2nd ed. New York: John Wiley and Sons.

[15] van de Hulst, H.C. 1981. *Light Scattering by Small Particles.* New York: Dover Publications.

[16] *http://www.nist.gov/index.html.*

[17] Horowitz, P. and Hill, W. 1989. *The Art of Electronics.* 2nd ed. Cambridge: Cambridge University Press.

# 3

# Human and Biomimetic Sensors

## CONTENTS

## 3.1   Chapter Overview

The sensors in our bodies are the smallest, most sensitive, and fastest response sensors of all. This chapter provides the description, pathway, and characteristics of each human sensor. These are found in our visual, gustatory, olfactory, auditory, vestibular, and somatic sensory systems. Finally, examples of biomimetic sensors, those designed to mimic human sensors, are described.

## 3.2   Human Sensors

The human body contains well over one-quarter of a *billion* sensory receptors. These range from single sensory neurons, such as pain receptors, to more complex sensory organs, such as in the eye [1], [2]. Each eye, for example, has 127 million rods and cones that sense light. Vision, hearing, taste, smell, touch, temperature, pain, itch, equilibrium, and proprioception are sensed at the conscious level. Blood pressure, blood glucose concentration, pH and oxygen content of blood, pH of cerebral spinal fluid, osmolarity of body fluids, internal body temperature, and extensions in lung and gastrointestinal tract muscles are sensed unconsciously.

All sensory receptors within the body act in a similar manner. A certain type and amount of energy stimulates the sensory receptor, as displayed in Figure 3.1. The receptor then transduces the stimulus into an electrical

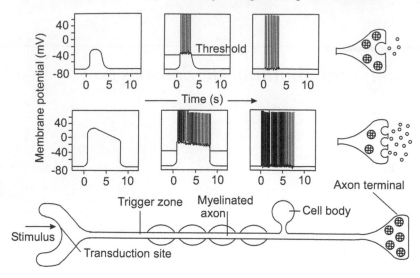

**FIGURE 3.1**

The sensory neuron and its response to a stimulus. Parts adapted from [1] and [2].

signal. This signal is a localized potential difference that is generated by a $Na^+$ and $K^+$ ion exchange through channels in the receptor membrane in the trigger zone region.

The potential difference can be either a graded potential or an action potential. Graded potentials have various amplitudes, each of which attenuates as it travels a short distance along the receptor. Graded potentials can sum if they occur within a short time interval (several ms). If the graded potential exceeds a threshold (a $\sim$20 mV increase with respect to approximately $-70$ mV resting potential) at or near the receptor cell's body (axon), an action potential is generated whose amplitude is fixed ($\sim$30 mV). This signal is short-lived ($\sim$5 ms) and travels along the receptor at speeds ranging from approximately 1 m/s to 70 m/s. This speed depends on the axon's diameter and whether or not it is insulated electrically (with myelin). The action potential eventually reaches the axon's terminus, which results in a release of chemicals (neurotransmitters) across an intercellular gap (synaptic cleft, $\sim$20 nm) to a neighboring neuron's dendrites. When an action potential is generated, the receptor is said to fire. Action potentials do not sum *within* a receptor.

When the stimulus leads to an action potential, the receptor fires at a frequency that is proportional to the intensity of the stimulus raised to a power. This is shown schematically in Figure 3.1. For example, in sensory nerve fibers, the percentage of the frequency of action potentials that occur for a touch stimulus with respect to the frequency for the maximum stimulus,

| Sense | Chemical Energy | Mechanical Energy | Electro-magnetic Energy | Thermal Energy |
|---|---|---|---|---|
| vision | - | - | √ | - |
| hearing | - | √ | - | - |
| taste | √ | - | - | - |
| smell | √ | - | - | - |
| equilibrium | - | √ | - | - |
| temperature | - | - | - | √ |
| touch | - | √ | - | - |
| proprioception | - | √ | - | - |
| nociception | √ | √ | - | √ |

**TABLE 3.1**
Classification of sensation according to type of stimulus energy. √ indicates type stimulus energy for that sense. Adapted from [1].

$R$, has been shown to be equal to $9.4S^{0.52}$, where $S$ is the percent of the magnitude of the maximum touch stimulus. The duration of a series of action potentials is proportional to the duration of the stimulus.

Stimulated sensory receptors fire in different manners. Some receptors (tonic receptors) fire rapidly at first and then decrease in firing rate to a constant value as long as the stimulus is present. These receptors monitor the steady state of a stimulus. Proprioceptors are examples of tonic receptors. Other receptors (phasic receptors) adapt and stop firing during stimulation. These fire rapidly at first and then cease to fire if the stimulus is maintained. These receptors, therefore, detect the introduction of the stimulus (actually, the positive derivative of the stimulus with respect to time). Smell receptors are phasic receptors.

The ending of the receptor that receives the stimulus can have free nerve endings, nerve endings enclosed in layers of tissue, or a specialized receptor cell that is connected to a sensory neuron through a synapse. The former two are termed neural receptors. The latter is called a non-neural receptor cell. Of the five special senses (vision, hearing, taste, smell, and equilibrium), only smell is detected directly by a neuron. The other four senses have specialized receptors.

A single sensory unit (a receptor with its dendrites or branches) senses a stimulus presented over an area. One sensory unit in the eye, for example, senses light over a surface area as large as 200 mm$^2$. The location of a stimulus can be closely identified through lateral inhibition. In this case, the response of neighboring secondary sensory neurons are inhibited by the primary neuron, which enhances perception of the stimulus location.

A stimulus can be characterized by its energy, location, intensity, and duration. Human sensory receptors respond to mechanical, chemical, thermal, and electromagnetic energy, as presented in Table 3.1. These are called mechanoreceptors, chemoreceptors, thermoreceptors, and photoreceptors, respectively. The stimuli of chemoreceptors are molecules that bind to the receptor site. Stimuli include oxygen, hydrogen (thus, pH), and more complex molecules. Mechanoreceptor stimuli are strain, vibration, acceleration, pressure, and sound. Light photons (electromagnetic energy) stimulate photoreceptors. Thermoreceptors are stimulated by temperature and changes in temperature. Each sensor description, its signal pathway, and sensor characteristics are presented in the following.

## 3.2.1   Vision

### Vision Sensor Description

There are two types of photoreceptors, cones and rods, with approximately 20 rods for every cone. Cones are used for high-acuity and color (photopic) vision, and rods for monochromatic, low-level nighttime (scotopic) vision. A rod has a radius of ~0.8 $\mu$m and a length of 30 $\mu$m [3]. A cone is of similar scale. Both types of receptors are contained within the retina, the sensory organ of the eye. Light is focused onto the retina, located at the back of the eye, by the lens. Cones and rods contain different photosensitive pigments that react to this light, which eventually produce action potentials.

### Vision Signal Pathway

The visual pathway through which electromagnetic energy is converted into an electrical signal is shown in Figure 3.2. Light encounters the surface of the retina after passing through the lens. It then travels through most of the retinal tissue to the outer segment where the membrane disks of the rods and cones are located. These disks contain the visual pigments. In rods, the pigment is rhodopsin (retinal, a form of vitamin A, bound to opsin, a protein). In cones, the process is similar, but with other visual pigments. When retinal is activated by light, it changes shape, causing it to unbind from opsin. *One* photon (an energy of only 358 zJ) activates *one* rhodopsin molecule. The conformational change in retinal causes the receptor to hyperpolarize and release a neurotransmitter across synapses to neighboring bipolar cells. These cells subsequently transmit to ganglion cells, each of which receives signals from bipolar cells covering an area of the retina.

### Vision Sensor Characteristics

Vision is most acute in the center of the visual receptor field, the fovea. This region is occupied mainly by cones, approximately 150 000/mm$^2$, spaced ~3 nm apart (3.6 arc seconds, based upon 0.29 mm/$°$). Approximately two-thirds of all cones are primarily sensitive to red, one-third to green, and a few to blue. The maximum number of rods per surface area,

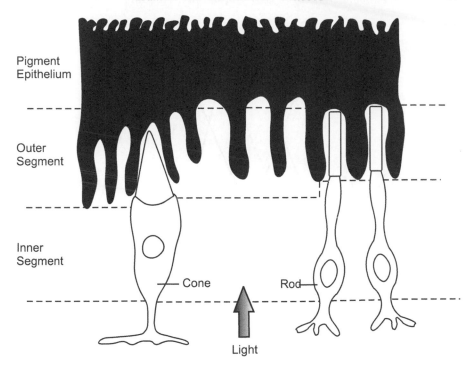

**FIGURE 3.2**
The pathway of phototransduction. Parts adapted from [1] and [2].

approximately 150 000/mm$^2$, occurs at $\pm 20°$ from the center. Mostly rods and far fewer cones are present up to $\pm \sim 80°$ from the fovea. The visual range (see Figure 3.3) extends from wavelengths of 397 nm (violet) to 723 nm (red), or photon energies from 1.8 eV to 3.1 eV. Vision is most sensitive at a wavelength of approximately 555 nm (2.23 eV), which is in the middle of the blue-through-green wavelength range. At that wavelength, as low as six photons of energy can be detected. The absolute intensity threshold for a dark-adapted eye is 10 nanolumens. The lower color threshold is approximately 20 nanolumens. Intensity changes as small as 1 % can be discriminated.

### 3.2.2 Taste and Smell

**Taste Sensor Description**

There are approximately 10 000 taste receptors (buds) on the human tongue. Each taste bud (see Figure 3.4) contains four different types of cells: basil, dark, intermediate, and light. The latter three are cells in various stages of development, with the light cell being the most mature. Taste cells are continually replaced, having a half-life of approximately 10 days.

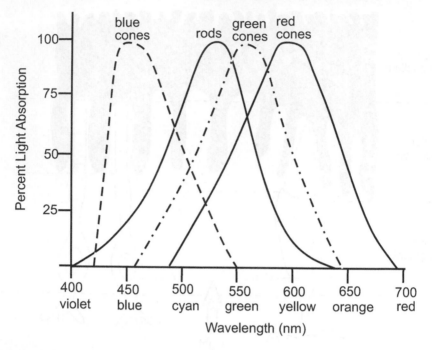

**FIGURE 3.3**
The visible spectrum and response curves for rods and cones. Parts adapted from [1] and [2].


Taste receptors are on the surface of the tongue and in the mucosa of the epiglottis, palate, and pharynx. There are three types of protuberances on the tongue (papillae), each containing a different number of taste buds (from approximately 5 to 100 taste buds per papilla). The sensory end of a taste bud has microvilli that are located within a taste pore. These act to increase the sensory surface area of a taste bud. The connectivity of taste cells with their nerve fibers is distributed. Each taste bud connects to approximately 50 nerve fibers, where each nerve fiber receives input from approximately 5 taste buds.

### Taste Signal Pathway

What is tasted (a tastant) first is dissolved with mucus and saliva. Remnant ligands of the tastant contact the microvilli of the taste bud within the taste pore. Different proteins, located on the outer surface of the taste cell, react with the ligands responsible for each of five different taste sensations: bitter, salty, sour, sweet, and umami. All five of these are sensed throughout the gustatory system. This chemical interaction opens selective ion channels, which leads to the depolarization of the cell. The resulting action potential releases either a neurotransmitter across a synapse (for salty and sour sen-

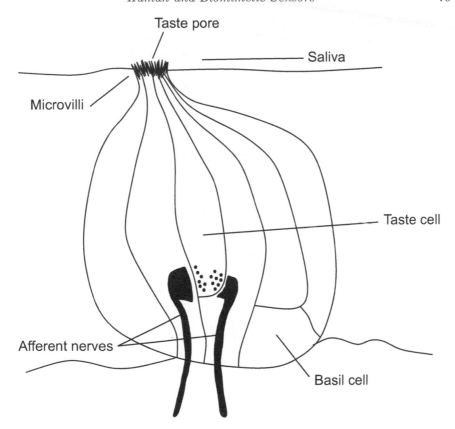

**FIGURE 3.4**
The taste bud. Sensory microvilli are located within the taste pore away from the taste pore surface. Parts adapted from [1] and [2].

sations) or ATP through a gap junction-like channel (for bitter, sweet, and umami sensations). The neurotransmitter contacts a primary sensory neuron of the gustatory afferent nerve (either the facial or the glosopharyngeal nerve). The ATP acts on gustatory neurons and neighboring cells. The signal then propagates via the solitary nerve tract to the thalamus and cortex.

### Taste Sensor Characteristics

The concentration thresholds for various substances that represent the different taste sensations are in the $\mu$mol/L to mmol/L range. The threshold for strychnine hydrochloride (representing bitter sensation) is 1.6 $\mu$mol/L. The threshold for glucose (representing sweet sensation) is 80 mmol/L. Similar to smell, a concentration change of approximately 30 % is necessary to detect a tastant concentration change.

### Smell Sensor Description

There are 10 to 20 million olfactory sensory neurons (see Figure 3.5), covering a $\sim5$ cm$^2$ surface area of the olfactory epithelium located on the roof of the nasal cavity. Each of these neurons expresses only one of 1000 different odorant receptors. The sensory end of the neuron has a thick dendrite, whose knobby end contains 10 to 20 cilia ($\sim0.1$ $\mu$m in diameter and $\sim2$ $\mu$m in length). These cilia are embedded in the nasal mucosa (a layer of mucus) and contain odorant receptor proteins.

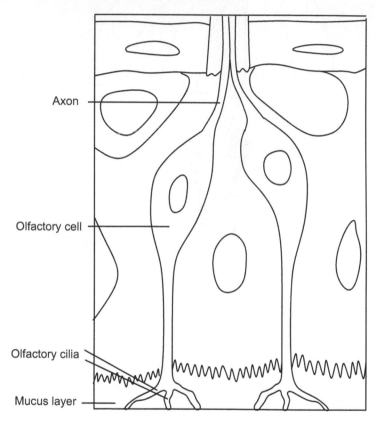

**FIGURE 3.5**

The olfactory cell. Parts adapted from [1] and [2].

### Smell Signal Pathway

An odor is sensed by first having its molecules dissolve and move in the nasal mucus layer, eventually reaching and binding to an odorant receptor protein on olfactory cilia (see Figure 3.5). This chemical binding activates the protein, which leads to the opening of ionic channels. This subsequently yields a graded receptor potential within the olfactory cell that develops into an action potential in the olfactory nerve. The signal transmission end of the neuron terminates in the olfactory bulb. The spatial organization and connections of neurons in the olfactory pathway produce a two-dimensional mapping in the olfactory bulb of each odorant. The additional dimensionality and connectivity helps to explain why 1 000 different odorant receptors can detect millions of different odors.

### Smell Sensor Characteristics

More than 10 000 different odors can be recognized by humans. The molecules responsible for this plethora of odors typically contain less than 20 carbon atoms, some with the same number of carbon atoms but different structural configurations. Olfactory sensory neurons are quite sensitive. In some cases, the smell of a single odorant molecule can be detected. Methyl mercaptan (in garlic) can be sensed at a concentration as low as 500 pg/L of air, while ethyl ether can be sensed at 6 mg/L of air. Odorant concentration changes, however, must be approximately 30 % for a human to detect a change.

## 3.2.3 Hearing and Equilibrium

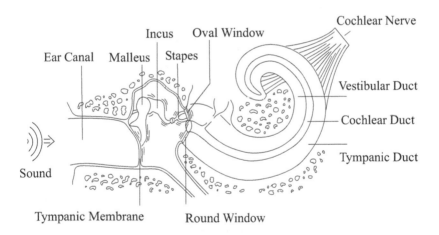

**FIGURE 3.6**
Components of the ear. Parts adapted from [1] and [2].

Both the sensors for hearing and for equilibrium are located within the ear, as depicted in Figure 3.6. The external section of the ear (the outer ear, or pinna, and the ear canal) functionally directs sound waves to the eardrum (the tympanic membrane). This membrane covers the entrance to the middle ear. The middle ear houses three bones (the malleus, which is connected to the eardrum, the incus, and the stapes, which is connected to the oval window of the inner ear). This system of bones transduces sound waves into minute mechanical vibrations. The middle ear is filled with air

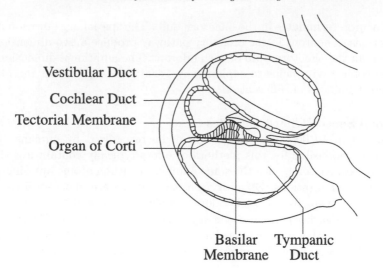

Vestibular Duct
Cochlear Duct
Tectorial Membrane
Organ of Corti

Basilar     Tympanic
Membrane    Duct

**FIGURE 3.7**
The cochlear duct, tympanic, and vestibular ducts. Parts adapted from [1] and [2].

that is provided by the collapsible eustachian tube, which is connected to the pharynx. The mechanical vibrations of the oval window's membrane set up pressure waves within the fluid contained in the cochlear duct of the inner ear. These waves eventually stimulate hearing sensory hair cells. Also within the inner ear is the fluid-filled vestibular apparatus. This contains the semicircular canals responsible for equilibrium.

**Auditory Sensor Description**
There are approximately 25 000 hair cell receptors inside each human cochlear duct (see Figure 3.7) that are the actual auditory sensors. These receptors (see Figure 3.8) are embedded in an epithelial layer containing primary sensory neurons and supporting cells, together constituting the organ of Corti. The basilar membrane supports this organ. The tectorial membrane overlays it. The sensory end of each hair cell has approximately 100 stiffened cilia (stereocilia) of increasing height. The longest (the kinocilium) extends across a fluid-filled gap and is embedded in the tectorial membrane. The shorter stereocilia extend into the fluid but do not touch the tectorial membrane.

**Auditory Signal Pathway**
As a pressure wave moves through the cochlear duct, it displaces both the basilar and tectorial membranes. This, in turn, moves both the kinocilium and the stereocilia. This moves protein bridges (tip links) between the stereocilia, which opens and closes ion channels, thereby producing a membrane

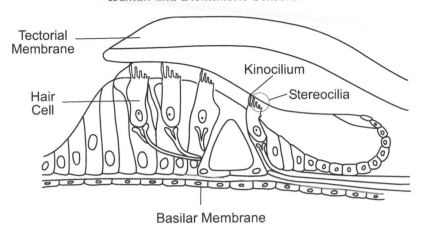

**FIGURE 3.8**
Auditory sensory hair cells and their supporting structure. Parts adapted from [1] and [2].

potential change in the hair cell. This change in membrane potential is proportional to the direction and degree of displacement of the stereocilia with respect to the kinocilium. Resulting action potentials of hair cells release neurotransmitters across synapses to nerve fibers of the cochlear (auditory) nerve. This information eventually reaches the auditory cortex.

### Auditory Sensor Characteristics
Sound can be characterized by its amplitude (loudness) and frequency (pitch). The minimum auditory amplitude threshold in humans is 20.4 $\mu$Pa of pressure (a slight whisper in a quiet room), corresponding to 0 dB by definition. The maximum pressure that potentially damages the cochlea is 10 million times higher, which is 200 Pa or 140 dB. The threshold varies with frequency, as shown in Figure 3.9. Sound frequencies audible to humans range from approximately 16 Hz to 28 kHz. A human can distinguish an average of about 2000 different frequencies. Changes in frequency as small as 0.03 % can be detected in the most sensitive frequency ranges. The structure of the cochlear duct establishes sensitivity to various frequencies. The duct spirals two and three-quarter revolutions inward over a length of ~35 mm. The width at its inlet (apex) is 500 $\mu$m. This width gradually decreases to 40 $\mu$m at its end (base). Pressure waves with 200 Hz frequencies give maximum displacement near the apex, while those with 20 kHz frequencies give maximum displacement near the base.

### Vestibular Sensor Description
Hair cell receptors, similar to those found in the auditory sensory system, help maintain equilibrium. These are located within the fluid-filled vestibular apparatus of the inner ear (see Figure 3.10). Two components

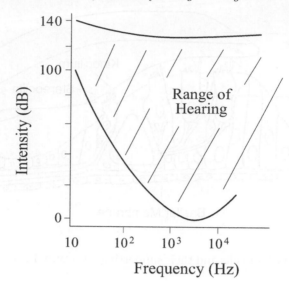

**FIGURE 3.9**
The auditory response curve of the human. Parts adapted from [1] and [2].

of the vestibular apparatus, the saccule and the utricle, sense linear acceleration and head position. These are positioned 90° with respect to each other. Three other components, the semicircular canals, which physically are oriented orthogonally to each other, sense rotational acceleration in three orthogonal directions. At the end of each canal is an enlarged chamber (the ampulla) that contains the sensory component. This is comprised of hair cells contained within a gelatinous mass, the cupula, which extends across the ampulla, as illustrated in Figure 3.11.

**Vestibular Signal Pathway**

Rotation is sensed via the semicircular canals when the head begins to rotate. The fluid inside it lags because of its inertia. This produces drag on the cupula and causes it to bend in the direction opposite of rotation. As the rotation continues, the cupula returns to its normal position within approximately 30 s. When rotation stops, the fluid continues to move and drag the cupula in the opposite direction. The sensory structures inside the saccule and utricle are somewhat different. Their hair cells are embedded also in a gelatinous mass (the otolith membrane), as shown in Figure 3.12. Crystals called otoliths lay atop this membrane and are bound to proteins on the membrane's surface. Gravitational displacement of the otoliths causes a movement of the membrane and its hair cells. Linear acceleration and head position changes with respect to the gravitational field are readily sensed. As in the auditory system, movements of a hairs cell's cilia in either one direction or another causes the cell to either hyperpolarize or depolarize.

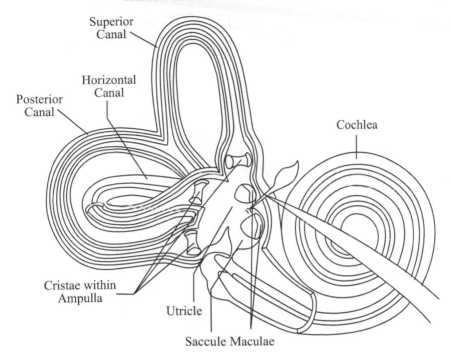

Superior
Canal

Horizontal
Canal

Posterior
Canal

Cochlea

Cristae within
Ampulla

Utricle

Saccule Maculae

**FIGURE 3.10**
The vestibular system. Parts adapted from [1] and [2].

This, in turn, leads to action potentials generated in the primary afferent neuron that is connected synaptically to the hair cell. Resultant signals travel via different neural pathways to various parts of the brain, including the cerebellum, reticular formation, and thalamus, as well as to the the spine and to nuclei that control eye movement.

### Vestibular Sensor Characteristics

There are approximately 38 000 neurons that are part of the human vestibular sensory system. Deflections of the cupula within the semicircular canals are proportional to velocity at which the head rotates, within rotational frequencies up to approximately 10 Hz. Hair cells constantly sense position and acceleration and thus produce high afferent neuron firing rates (over 1 million per second). This steady-state firing behavior allows for small changes to be sensed. Rotational accelerations, for example, as low as $0.1°/s^2$ can be sensed. Thresholds of discriminating linear accelerations in the interaural (left-right) and dorsoventral (up-down) directions are $0.06$ m/s$^2$ and $0.10$ m/s$^2$, respectively [4].

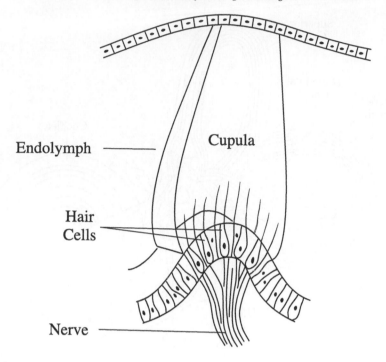

**FIGURE 3.11**
The cupula and its sensory hair cells. Parts adapted from [1] and [2].

### 3.2.4  Somatic

The somatosensors include those responsible for touch, proprioception, nociception (itch and pain), and temperature. They are located within the skin, viscera, muscles, tendons, and joints. Some of these sense directly from the dendritic nerve endings of primary afferent neurons (touch and temperature). Others sense with non-neural cells that are connected synaptically to primary afferent neurons (proprioception and nociception).

#### Touch Sensor Description

The sensors for touch, including pressure and vibration, are primary afferent nerve endings. These include Meissner corpuscles, Merkel receptors, Pacinian corpuscles, and Ruffini corpuscles (see Figure 3.13). Meissner corpuscles and Merkel receptors are located near the surface of the skin. The Meissner corpuscle, which is composed of dendrites contained within connective tissue, senses changes in texture and slow vibrations. The Merkel receptor, which has expanded dendrites, senses touch and sustained pressure. Pacinian corpuscles and Ruffini corpuscles are located in deeper layers of the skin. The Pacinian corpuscle, which consists of a single nerve ending contained within connective tissue, senses deep pressure and fast vibration.

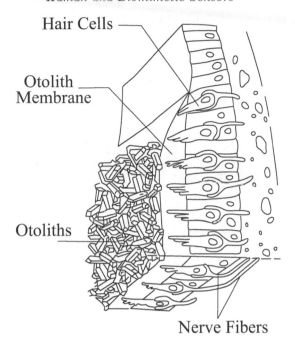

Hair Cells

Otolith
Membrane

Otoliths

Nerve Fibers

**FIGURE 3.12**
The otolith membrane with its crystals. Parts adapted from [1] and [2].

The Ruffini corpuscle, which has expansive nerve endings in a capsule of tissue, senses sustained pressure. Each sensor has a different sensory area, with the Meissner corpuscle having the smallest.

### Touch Signal Pathway
Action potentials are generated in touch sensor neurons by the opening of ion channels in their membranes. This opening is caused mechanically, by the compression of tissue connected to the nerve endings. The primary afferent neurons of touch sensors synapse in the medulla of the brain with secondary neurons. These neurons then connect to the thalamus.

### Touch Sensor Characteristics
Meissner corpuscles sense vibration from approximately 5 Hz to 40 Hz and Pacinian corpuscles from approximately 60 Hz to 300 Hz. The Pacinian corpuscle is optimally sensitive to 250 Hz vibrations, which corresponds to moving a finger across textures with wavelengths less than ~200 $\mu$m.

### Temperature Sensor Description
Temperature sensors are free nerve endings located near the surface of the skin (see Figure 3.13). They are classified as either cold or warmth

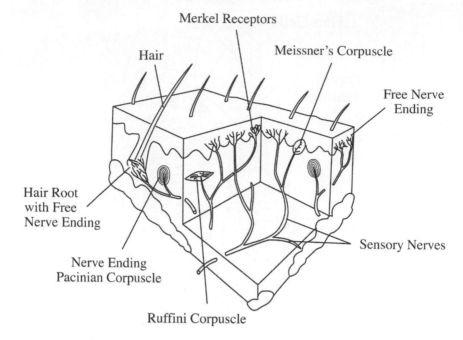

**FIGURE 3.13**
Touch and temperature sensors in the skin. Parts adapted from [1] and [2].

receptors. There are some cold sensors within the body, such as in the cornea and bladder. There are distinct cold- and heat-sensitive areas on the skin, with at least five times more cold-sensitive spots. Local temperature controls the opening of ion channels in these sensors' membranes, leading to action potentials generated at a certain frequency. As the temperature of an area changes, the action potential frequency changes.

### Temperature Signal Pathway
The signals from temperature sensors travel to secondary neurons in the spinal cord. These travel to the brain and terminate in the thalamus. A tertiary neuron then transmits the signal to the somatosensory cortex.

### Temperature Sensor Characteristics
Cold sensors respond to temperatures from ~10 °C to ~40 °C. From ~10 °C to ~24 °C, their firing frequency increases with temperature. From ~24 °C to ~40 °C, it decreases. Warmth sensors activate at ~30 °C and have an increased firing frequency up to ~46 °C.

### Nociception Sensor Description
Nociceptors, which sense pain and itching, are the dendritic fiber endings of primary afferent neurons. They respond to either chemical, mechanical,

or thermal stimuli. Each of these stimuli affect the receptor by opening ion channels in its membrane.

### Nociception Signal Pathway

The signal pathway of nociceptors to the brain is similar to that of the temperature sensors. Additionally, nociceptors are connected directly to efferent neurons in the spinal cord such that an immediate withdrawal response can be initiated upon tissue insult, as in touching a very hot surface.

### Nociception Sensor Characteristics

The perception of pain is subjective and varies from individual to individual. Thus, specific thresholds usually are not given. The pressure pain threshold, averaged over 60 subjects, obtained by applying a force to the tip of the middle finger using a dolorimeter was approximately 75 N/cm$^2$ (4.75 lbf over a 0.6-cm-diameter hemisphere) [5].

### Proprioception Sensor Description

Proprioceptors provide a "sense of self," sensing the spatial position of our limbs, their movement, and the state of their muscles' tension. They include muscle spindles (see Figure 3.14), Golgi tendon organs (see Figure 3.15), and joint receptors. They sense changes in muscle length, muscle tension, and relative position of bones, respectively. All three sensors are primary afferent nerve endings. Muscle spindles, present within muscle, consist of connective tissue, intrafusal muscle fibers, efferent nerve endings, and two kinds of sensory nerve endings. One kind of sensory ending (group 1a primary) monitors the speed of change in muscle length. Another kind (type

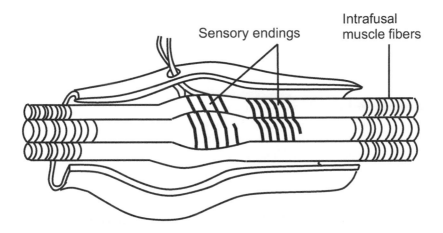

**FIGURE 3.14**
The muscle spindle. Parts adapted from [1] and [2].

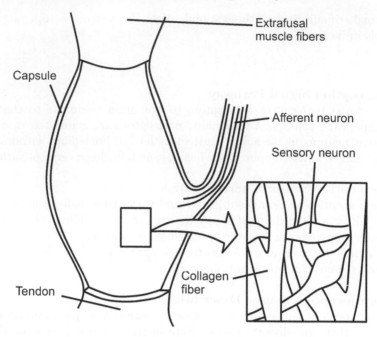

**FIGURE 3.15**
The Golgi tendon organ. Parts adapted from [1] and [2].

II secondary) monitors the steady-state muscle length along with the primary nerve ending. Spindles are arranged in parallel within a muscle. These stretch receptors sense muscle length and changes in length. Golgi tendon organs connect at one end to muscle fibers and at the other end to tendons. Thus, they are in series with the extrafusal muscle fibers. They comprise connective tissue, collagen fibers, and afferent nerve endings. They sense the contraction of muscle but not its extension. The afferent nerve endings of joint receptors respond to changes in the relative position of bones that are linked by flexible joints.

**Proprioception Signal Pathway**
The muscle spindle is stretched when a muscle lengthens. This causes contractile proteins, located at the ends of the intrafusal fibers, mechanically to open ion channels in the dendrites, generating an action potential in its neuron. Collagen fibers within the Golgi tendon organ, in response to muscle contraction, contract and squeeze sensory afferent nerve dendrites, mechanically opening ion channels. A similar mechanical opening of ion channels occurs in joint receptors. All of these proprioceptors connect synaptically in the spinal cord with afferent pathway neurons that travel to the brain's central nervous system and either directly or indirectly with somatic motor neurons. The direct connection to somatic motor neurons is involved in

the monosynaptic stretch reflex such as the knee-jerk reflex. The indirect connection is utilized in movement as a response to a painful stimulus.

### Proprioception Sensor Characteristics

Each muscle spindle contains approximately eight intrafusal muscle fibers and about the same number of nerve endings, as depicted in Figure 3.14. One small finger muscle in a newborn has approximately 50 muscle spindles. Each Golgi tendon organ has a single (type 1b) sensory nerve ending and connects to about 20 extrafusal muscle fibers.

## 3.3 Biomimetic Sensors

The field of biomimetics studies nature, with the goal of mimicking its methods, mechanisms, and processes [6], [7], [8]. Notable progress has been made recently in biomimetics, particularly in mimicking biological materials and the modes of motion, including flight, walking, and hopping, of various animals and insects.

Biomimetic sensor technology has made recent significant advances. Biomimetic sensors have been developed that sense mechanical variables, such as strain, deformation, and position, and have voltage outputs. An extensive review by Stroble *et al.* [9] cites 33 different biomimetic sensors whose descriptions have been published prior to 2009. These sensors encompass 12 mechanical, seven electric, six acoustic, five chemical, two optical, and two thermal. The types of receptors mimicked include mechano-, photo-, chemo-, thermo-, proprio-, electro-, and magneto-receptors.

Most biomimetic sensors developed thus far do not directly mimic human sensors. Partial mimicry of some human sensory systems has been accomplished. This includes the cochlea, the taste bud, the olfactory epithelium, the eye, muscle proprioceptors, Pacinian corpuscles, and the skin. The actual sensors in such systems are not yet the same as human sensors. The human primary sensory neuron converts some form of received stimulus energy directly into an electrical signal (a local ion exchange across the neuron's membrane). Further, it has a higher sensitivity, better stimulus-to-signal conversion efficiency, and the most direct and physically smallest path from sensation to signal.

In the following, three relatively recent biomimetic systems are highlighted. The first system mimics the navigational system of a fly, the second a mechano-sensing hair, and the third a human tactile sensor. These systems are described to elucidate the level of operational sophistication and size that has been achieved thus far.

### Multi-sensor Flight System

A group biomimetic sensors have been developed for the navigational system of a micromechanical flying insect [10]. This suite includes a photodiode array to sense light intensity, piezo-actuated vibrating structures to sense rotational velocities, elemental motion detectors to sense direction, and magnetic field sensors to sense magnetic field intensity. The light intensity sensor consists of four infrared photodiodes arranged in a pyramidal array. The output voltages, each measured across a resistor in parallel with a photodiode, are combined as differences to yield two outputs. These outputs, after calibrating the array, provide an estimate of the distance from the light source to the array. The entire array is 125 mm$^3$ and has a mass of 150 mg. Rotational velocities are sensed using flat beams, twisted and with masses on their ends, and with strain gages mounted on each flat side in the twist region of the beam. These beams are vibrated using a piezoelectric actuator, generating a Coriolis force, sensed using the strain gages, that is proportional to the rotational velocity. The motion detectors operate in pairs whose conditioned signals, which are proportional to light intensity, are correlated to yield direction. Three magnetic field sensors are connected in a three-dimensional array, ~10 mm in dimension. Each loop is a conducting U-shaped cantilever through which current flows, with a strain gage mounted at the cantilever base. The terrestrial magnetic field induces a Lorenz force on the cantilever, which is measured using the strain gage system. This output is used to determine the direction of the array with respect to the terrestrial magnetic field.

### Acoustic Sensor

A hair flow-sensor developed for acoustic sensing [11]. The hair-sensors are fabricated by forming silicon nitride suspended membranes and using polymer processing for hair fabrication. A highly conductive silicon wafer serves as the substrate. A conducting chromium layer is deposited over an insulating silicon nitride layer that is placed on top of the substrate. These electrodes form a variable capacitor. An artificial hair (~1 mm long and ~ 500 $\mu$m diameter), developed using SU-8 photoresist material, is formed on the insulating layer. This artificial hair extends through, and is normal to, the chromium layer. The capacitance is altered by the movement of the hair, which slightly displaces the chromium conducting layer, thereby effectively changing the distance between the two conducting plates. The capacitance is converted into a voltage after the signal from the sensor is amplified and demodulated by a charge amplifier and synchronous detector. This system can detect flow velocities from ~ 0.1 m/s to ~ 1 m/s.

### Tactile Sensor

A tactile sensor, the size of a human fingertip, has been developed to sense the magnitude of vibrations induced by the motion of textured surfaces over the sensor [12]. This sensor mimics the human Pacinian corpuscle, which has low spatial resolution yet can discern textures smaller than

~200 $\mu$m. The sensor consists of a rigidly mounted micro-electro-mechanical force sensor. That sensor is covered with an elastic cap that represents the fingertip skin. This is mounted on a double-cantilever system containing capacitive position sensors. The outputs of the complete system are voltages corresponding to the sensed normal and tangential forces as a function of time. These signals can be transformed into a record of sensed pressure versus time.

## 3.4   Problems

1. A touch stimulus is applied to the tip of an index finger. The firing of a sensory nerve fiber in that region is monitored. At a maximum touch stimulus of 8 N, the nerve fiber fires at a frequency of 95 Hz. Determine the firing frequency (in Hz) upon a stimulus application of (a) 4 N and (b) 2 N.

2. A cell maintains a potential difference (outside minus inside) of 120 mV across its membrane under standard and equilibrium conditions. For this situation, the electrochemical activity of the ion that is responsible for this potential difference is constant. This activity, as determined from the Nernst equation 2.84, equals $[C] \exp(zFV/\mathcal{R}T)$, in which $[C]$ is the concentration, $z$ the number of net electrical charges of the ion, and $V$ the potential. Assuming that sodium ions only participate in this passive process, determine (a) the ratio of the outside-to-inside sodium ion concentrations. Next, if chloride ions were participating instead of sodium ions, determine (b) the ratio of the outside-to-inside chloride ion concentrations.

3. The internal quantum efficiency of a CCD (charge-coupled device) is defined as the fraction of the photons that enter the CCD array that produce a detected photo-electron. This fraction is as large as 0.85 for some CCDs across the full visible spectrum. Compare this efficiency to that of one rhodopsin molecule on a rod in the eye.

4. (a) What physical events occur when a primary sensor neuron receives stimulus energy that are common to all types of human primary sensors? (b) What physical events are different? (c) For each of the senses, describe the specific manner in which the sensory neuron's ion channels are opened when the neuron receives stimulus energy.

5. Determine (a) the approximate frequency (in Hz) at which a human has the greatest range of hearing. Then, (b) determine that range (in $\mu$Pa).

# Bibliography

[1] Barrett, K.E., Barman, S.M., Boitano, S., and Brooks, H.L. 2010. *Ganong's Review of Medical Physiology.* 23rd ed. New York: McGraw-Hill, Inc.

[2] Silverthorn, D.U. 2010. *Human Physiology An Integrated Approach.* 5th ed. New York: Pearson/Benjamin Cummings.

[3] van Kuijk, F.J., Lewis, J.W., Buck, P., Parker, K.R., and Kliger, D.S. 1991. Spectrophotopic Quantitation of Rhodopsin in the Human Retina. *Invest. Opth. and Visual Sci.*, 32, 1962–1967.

[4] MacNeilage, P.R., Banks, M.S., DeAngelis, G.C., and Angelaki, D.E. 2010. Vestibular Heading Discrimination and Sensitivity to Linear Acceleration in Head and World Coordinates. *J. Neuroscience*, 30, 9084–9094.

[5] Ozcan, A., Tulum, Z., Pinar, L., and Baskurt, F. 2004. Comparison of Pressure Pain Threshold, grip Strength, Dexterity and Touch Pressure of Dominant and Non-ominant Hands within and between Right- and Left-Handed Subjects. *J. Korean Med. Sci.*, 19, 874–878.

[6] Schmitt, O.H. 1969. Some Interesting and Useful Biomimetic Tranforms. International Biophysics Conference, Boston, MA.

[7] Bar-Cohen, Y. 2006. Biomimetics — Using Nature to Inspire Human Innovation. *Bioinsp. Biomim.*, 1, P1–P12.

[8] Toko, K. 2005. *Biomimetic Sensor Technology.* Cambridge: Cambridge University Press.

[9] Stroble, J.K., Stone, R.B., and Watlins, S.E. 2009. An Overview of Biomimetic Sensor Technology. *Sensor Review*, 29, 112–119.

[10] Wu, W.-C., Schenato, L., Wood, R., and Fearing, R.S. 2003. Biomimetic Sensor Suite for Flight Control of a Micromechanical Flying Insect: Design and Experimental Results. *Proceedings of the IEEE International Conference on Robotics and Automation*, 1, 1146–1151.

[11] Krijnen, G.J.M., Dijkstra, M., van Baar, J.J., Shankar, S.S., Kuipers, W.J., de Boer, R.J.H., Altpeter, D., Lammerink, T.S.J., and Wiegerink, R. 2006. MEMS Based Hair Flow-Sensors as Model Systems for Acoustic Perception Studies, *Nanotechnology*, 17, S84–S89.

[12] Scheibert, J., Leurent, S., Prevost, A., and Debregeas, G. 2009. The Role of Fingerprints in the Coding of Tactile Information Probed with a Biomimetic Sensor. *Science*, 323, 1503–1506.

# *Index*